METHODS FOR PROTEIN ANALYSIS

Robert A. Copeland, PhD
The DuPont Merck Pharmaceutical Company
Experimental Station
P.O. Box 80400
Wilmington, DE 19880-0400

METHODS FOR PROTEIN ANALYSIS

A Practical Guide to Laboratory Protocols

ROBERT A. COPELAND

SPRINGER-SCIENCE+BUSINESS MEDIA, B.V.

Library of Congress Cataloging in Publication Data

Copeland, Robert Allen.
Methods of protein analysis : a practical guide to laboratory protocols / Robert A. Copeland.
p. cm.

Includes bibliographical references.
ISBN 978-1-4757-1507-1 ISBN 978-1-4757-1505-7 (eBook)
DOI 10.1007/978-1-4757-1505-7

1. Proteins—Analysis—Laboratory manuals. I. Title.
QP551.C725 1993
547.7'5'028—dc20 93-6934
CIP

British Library Cataloguing in Publication Data also available.

Dedication

This book is dedicated, with all my love always, to Nancy

Preface

With the revolution that has occurred in Molecular Biology over the past twenty years, an exponential growth has occurred in the number of laboratories and researchers that now find themselves working in the field of protein science. The tools of recombinant DNA methods have been largely responsible for this intellectual growth spurt, and many excellent texts have appeared on the subject of these technologies to keep up with the ever growing demands of the research community. Of particular value to the bench scientists are texts aimed at providing detailed laboratory protocols, rather than details of the theoretical basis for many of these methods. Thus, the series of laboratory manuals that have emerged from the Cold Spring Harbor Laboratories, for example, have been especially well received by the Molecular Biology community. Surprisingly, however, until very recently there have not been comparable texts devoted to the analysis of proteins.

In the text that follows, I have tried to fill this perceived need by the protein community. This field is so diverse, however, that any attempt at a comprehensive treatment of protein science would require a multi-volume collection. I therefore have made no attempt to be comprehensive here. Rather, I have chosen those methods that I consider most relevant to the generalist in protein science. Some methods that are more specialized, like amino acid sequence analysis, have been given only a cursory treatment here. This is in recognition of the fact that most general protein laboratories will not participate directly in these studies, but are more likely to submit samples for sequence analysis to specialized laboratories. Other methods, such as analytical ultracentrifugation, that provide a wealth of information of protein hydrodynamic properties have not been covered in the interest of brevity.

The methods covered here are those that can reasonably be performed in a standard protein laboratory, without specialized equipment or expertise. The techniques are presented in a "cookbook" fashion, emphasizing the how-to approach to science, and de-emphasizing, as much as possible, the theoretical basis for these methods. This bias is based on my personal belief that this type of treatment is most greatly needed in industrial, academic, and government laboratories at present.

Robert A. Copeland

Wilmington, Delaware
January 20, 1993

Acknowledgments

I am grateful to the many colleagues with whom I have collaborated over the years, and who have added to my knowledge of protein science. I would especially like to thank Professor Howard Tager of the University of Chicago for all his help, friendship, and advice. I would also like to express my appreciation to my students at the University of Chicago for their good humor and hard work during the writing of this manuscript. I am grateful to the Department of Chemistry and Biochemistry of the University of Delaware, which graciously provided me a place to work during the final preparation of this work. Finally, I would like to thank my family—Nancy, Lindsey, and Amanda—for putting up with me during the numerous crises that arose while this book was being written.

1

Introduction to Protein Structure

Proteins (from the Greek word *protos*, meaning first) occur ubiquitously throughout all organisms, where they participate in a wide variety of biological functions. Proteins provide the structural rigidity *and* flexibility associated with external body parts such as skin, hair, and the exoskeletons of arthropods. By changing their structure they provide the basis for biological motions, such as muscle contraction. A large class of proteins, the enzymes, serve as biological catalysts, enhancing greatly the rates of chemical reactions that are vital to life (Copeland, 1992).

How do proteins accomplish this great diversity of function? The answer to this lies in the structure and the structural variation accessible to proteins. In this chapter we will describe the structural components that make up proteins, the amino acids, and regular patterns of three dimensional arrangements of these groups that are commonly found in proteins. Ultimately it is the chemical reactivities of the individual amino acids and their arrangement in space that uniquely identify the active conformation of a protein and determine its functional activity.

THE AMINO ACIDS

Nature has selected twenty compounds to use as the building blocks for all proteins and peptides. These twenty molecules share the common structure shown in figure 1.1, and are collectively known as the natural amino acids. At neutral pH all of the amino acids are zwitterionic, having both a positively charged amino group and a negatively charged carboxylate group (hence the name amino acid). Intervening between these

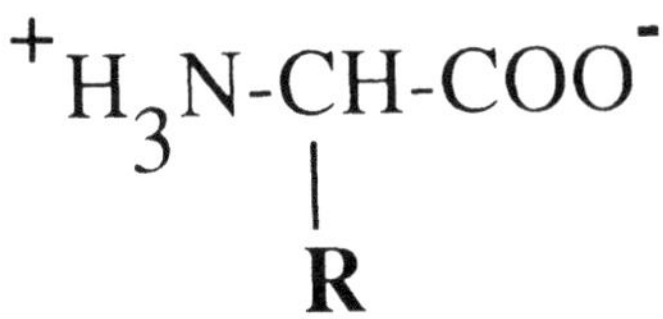

Figure 1.1 The general structure of an amino acid in its zwitterionic form.

charged functional groups is a single tetrahedral carbon atom, referred to as the alpha carbon. What distinguishes the amino acids from one another, both in terms of structure and chemical reactivity, is the identity of the substituent linked to the alpha carbon, which is denoted by R in figure 1.1 and is commonly referred to as the *amino acid side chain*. The side chain can be as simple as a hydrogen, in the case of glycine, or as complex as a fused bicyclic ring, in the case of tryptophan. Table 1.1 lists the twenty naturally occurring amino acids and illustrates the structures of their side chains. Also included in table 1.1 are the three-letter and one-letter abbreviations for the amino acids that are commonly used in the protein literature. The pK_a values for the amino, carboxyl, and side chain groups of each amino acid are also listed here. A quick perusal of this table will convince the reader that the pK_a values for the amino and carboxyl groups of all the amino acids are similar, hovering around 9.5 and 2.0, respectively. The pK_a values for the side chains, on the other hand, vary considerably (Dawson et al., 1969).

PEPTIDE FORMATION AND SEQUENCE NOMENCLATURE

Two or more amino acids can be covalently linked through a condensation reaction leading to an amide bond:

$$NH_3^+-CH(R_1)-COO^- + NH_3^+-CH(R_2)-COO^- \rightarrow$$
$$NH_3^+-CH(R_1)-C(O)-NH-CH(R_2)-COO^- + H_2O$$

The amide linkage formed in this process is called a *peptide bond*. When several amino acids are linked together in this fashion, the resulting polymer is called a *polypeptide*, and the individual amino acids along the polymer chain are called *residues*. One may ask what the difference is between a polypeptide and a protein. The difference is largely semantic. All proteins are, by definition, polypeptides. The term polypeptide is usually reserved for small polymers of, say, 50 or fewer amino acids.

Table 1.1 The structures of the naturally occurring amino acids found in proteins. This table is adapted from Dawson et al. (1969), with permission.

Amino acid (three- and one-letter codes, M_r)	Side chain R in $RCH(NH_3^+)CO_2^-$	pK_a's[a]
Glycine (Gly, G, 75)	H—	2.35, 9.78
Alanine (Ala, A, 89)	CH_3—	2.35, 9.87
Valine (Val, V, 117)	H_3C CH— H_3C	2.29, 9.74
Leucine (Leu, L, 131)	H_3C $CHCH_2$— H_3C	2.33, 9.74
Isoleucine (Ile, I, 131)	CH_3CH_2 CH— CH_3	2.32, 9.76
Phenylalanine (Phe, F, 165)	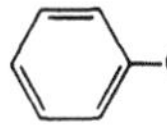CH_2—	2.16, 9.18
Tyrosine (Tyr, Y, 181)	HO—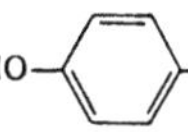CH_2—	2.20, 9.11, 10.13
Tryptophan (Trp, W, 204)	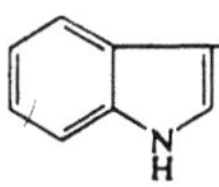CH_2—	2.43, 9.44
Serine (Ser, S, 105)	$HOCH_2$—	2.19, 9.21
Threonine (Thr, T, 119)	HO CH— H_3C	2.09, 9.11
Cysteine (Cys, C, 121)	$HSCH_2$—	1.92, 8.35, 10.46
Methionine (Met, M, 149'	$CH_3SCH_2CH_2$—	2.13, 9.28
Asparagine (Asn, N, 132)	$H_2NC(=O)CH_2$—	2.1, 8.84
Glutamine (Gln, Q, 146)	$H_2NC(=O)CH_2CH_2$—	2.17, 9.13
Aspartic acid (Asp, D, 133)	$^-O_2CCH_2$—	1.99, 3.90, 9.90
Glutamic acid (Glu, E, 147)	$^-O_2CCH_2CH_2$—	2.10, 4.07, 9.47
Lysine (Lys, K, 146)	$H_3N^+(CH_2)_4$—	2.16, 9.18, 10.79
Argininc (Arg, R, 174)	H_2N^+ C—$NH(CH_2)_3$— H_2N	1.82, 8.99, 12.48
Histidine (His, H, 155)	CH_2—	1.80, 6.04, 9.33
Proline (Pro, P, 115)	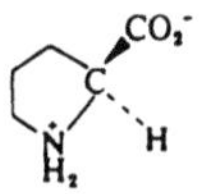	1.95, 10.64

A protein or polypeptide is uniquely defined by its amino acid composition, and by the order in which the amino acids occur along the linear chain of the polymer. This information is referred to as the *primary structure* or *amino acid sequence* of the protein. Note that no matter how many amino acids we string together to form a protein, we are always left with one terminal amino acid retaining its positively charged amino group, and the other terminal amino acid retaining its negatively charged carboxyl group. These residues are denoted the amino or N-terminus, and the carboxyl or C-terminus, respectively.

The individual amino acids within a polypeptide are indexed numerically in sequential order. The amino terminal residue is always designated number 1; one then continues numbering the residues that follow in ascending numerical order, ending with the carboxyl terminus.

> When numbering the amino acids of a protein sequence, one always begins at the N-terminus and ends with the C-terminal amino acid.

This sequence nomenclature is illustrated in figure 1.2 for a pentapeptide.

PROPERTIES OF THE PEPTIDE BOND

When x-ray diffraction methods were first applied to small molecular weight amides, such as N-methyl acetamide, it was soon realized that the bond lengths observed in the crystal structures could not be accounted for by a typical carbonyl double bond (C=O) and a carbon-nitrogen single bond. Instead, both the carbonyl and carbon-nitrogen bond distances were intermediate between the known double and single bond distances for related compounds. To explain these results, delocalization of the π electron cloud over the triatomic O-C-N system is invoked (Creighton, 1984). This can be thought of in terms of the two resonance structures illustrated in figure 1.3.

Based on the crystallographically measured bond distances, the C-N

1 2 3 4 5

$^{+}NH_3$-Arg-His-Cys-Lys-His-COO^{-}

Figure 1.2 Sequence numbering system for proteins and polypeptides. Numbering begins with the N-terminal amino acid residue and proceeds in ascending sequential order ending with the C-terminal residue.

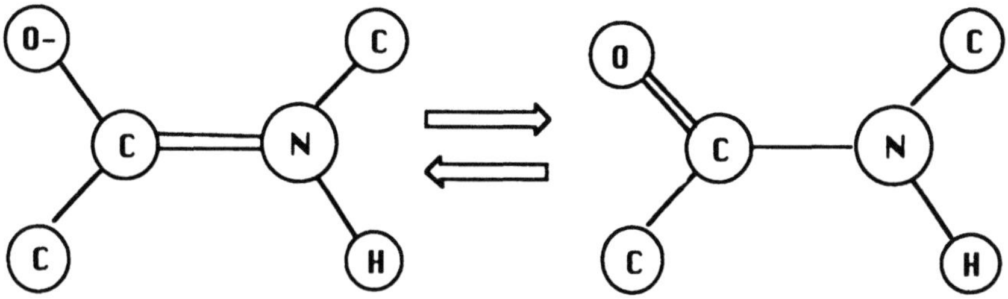

Figure 1.3 Resonance structures of the amide bonds of peptides.

bond appears to contain about 40% double bond character, and the C-O bond about 60% double bond character (Creighton, 1984). The significant π character of the C-N bond results in about a 20 kcal/mol resonance energy stabilization, and severe restriction of rotation about the C-N axis. Thus two peptide bond configurations are possible, cis and trans, with respect to the alpha carbons of the adjacent residues forming the peptide bond (see figure 1.4). In principle one could have polypeptides containing cis, trans, or both peptide bond isomers. However, the cis configuration leads to destructive non-bonding interactions that thermodynamically disfavor this conformation over the trans form by as much as three orders of magnitude. The vast majority of peptide bonds found in nature therefore occur in the trans configuration. One exception to this is prolyl peptide bonds. Occasionally, cis prolyl bonds have been observed in protein crystal structures, but even this is a rare event (Creighton, 1984).

Although rotation about the C-N bond is restricted, rotations about the C_α-N and C_α-C bonds can occur freely. The steric bulk of the amino acid side chain will, however, restrict to some degree the rotations about

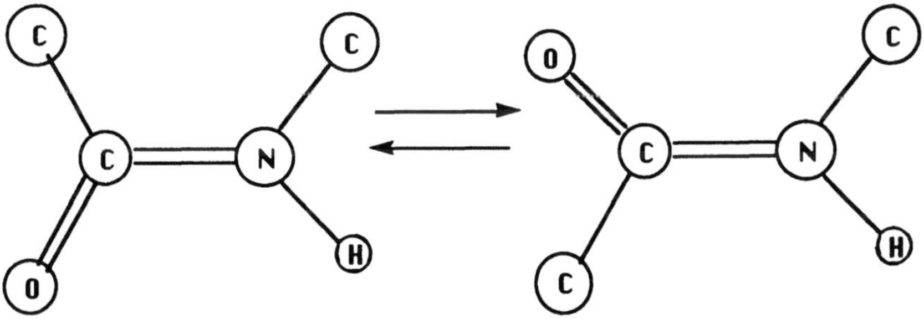

Figure 1.4 Cis and trans isomers of the amide bond. The trans form, shown on the left in this figure, is the thermodynamically favored form for all amide bonds found in proteins and polypeptides.

these bonds. One can define two dihedral angles to describe the orientation of peptide atoms with respect to these rotations, ψ and ϕ. These angles are illustrated in figure 1.5 for one peptide bond within a protein. In the late 1960's Ramachandran and coworkers surveyed the ψ and ϕ angles observed for amino acid residues within the crystal structures of proteins (Ramachandran and Sasisekhoran, 1968). Figure 1.6 illustrates the type of results one sees for such a survey for amino acids other than glycine. For glycine, the small size of the side chain (a proton) allows this amino acid greater freedom to survey extended ψ, ϕ space. What is most obvious from this plot is that the ψ and ϕ angles cluster around two sets of values. These two regions of high density correspond to the ψ and ϕ angles associated with two commonly occurring regular structural motifs that are found within proteins: the right-handed alpha helix, and the beta pleated sheet. These structural motifs are examples of protein *secondary structure,* an important aspect of the overall conformation of any protein (Branden and Tooze, 1991).

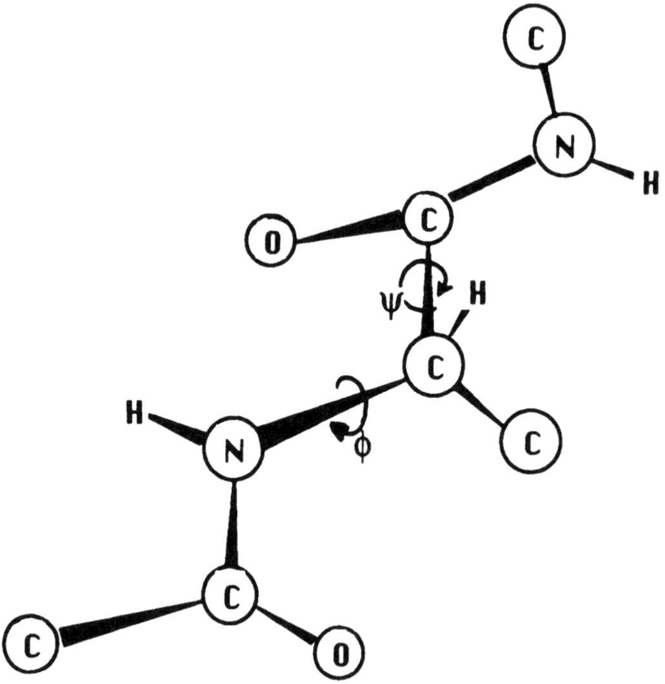

Figure 1.5 The dihedral angles of rotation for amino acids known as the Ramachandran angles, psi and phi.

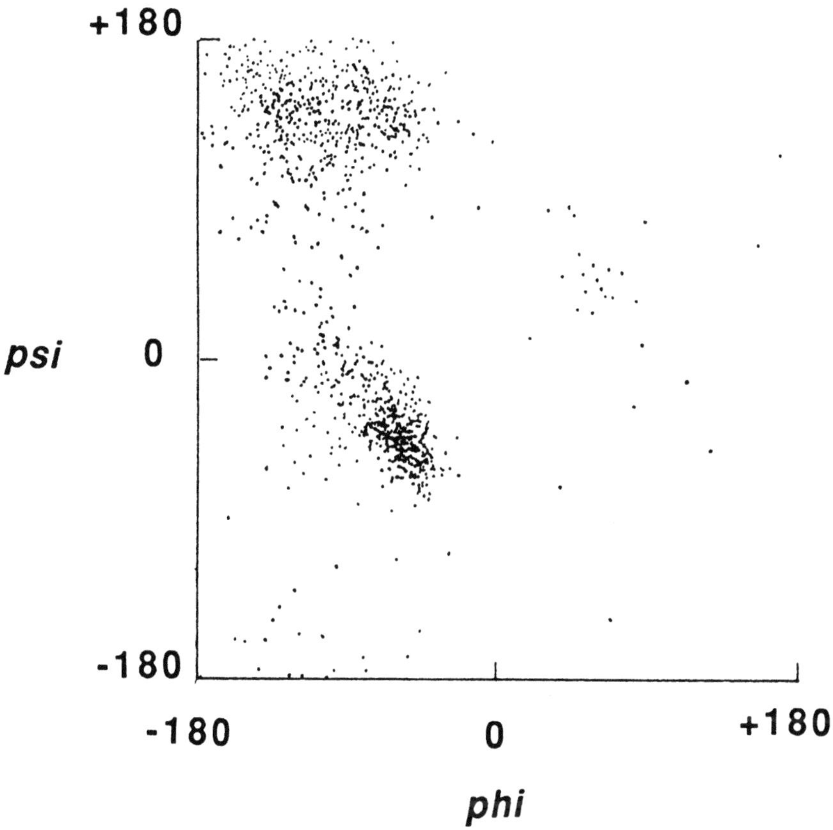

Figure 1.6 Ramachandran plot for the amino acid alanine illustrating the allowed combinations of psi and phi angles for this residue within proteins. Similar patterns are observed for all of the naturally occurring amino acids except for glycine. Because of the small size of the side chain of glycine (a proton), a wider range of allowed angle combinations is observed for this amino acid.

PROTEIN SECONDARY STRUCTURE

In the crystal structures of proteins one commonly finds regions of repeating structural patterns known as secondary structure. The two most commonly found secondary structures in globular proteins are the right-handed alpha helix, and the beta pleated sheet, which are illustrated in figure 1.7 (Pauling, 1960). These structural motifs are stabilized mainly by interamide hydrogen bonding. In the right-handed alpha helix, hydrogen bonding occurs between the backbone carbonyl of residue i and the nitrogenous proton of residue i+3 along the polypeptide chain. This pattern of hydrogen bonding confers a very specific and regular structure to the alpha helix. Each turn of the helix requires 3.6 residues with a translation along the helical axis of 1.5 Å per residue, or 5.4 Å per turn.

The beta pleated sheet represents a fully extended configuration of the polypeptide backbone, in which hydrogen bonding occurs between residues on adjacent strands of the polypeptide. Note that the two strands forming the beta sheet could be two sections of the same contiguous polypeptide, or form the interface between two separate polypeptides. If we imagine a beta sheet within the plane of this page, both strands of the sheet could run in the same direction, for instance from C-terminus at the top of the page to N-terminus at the bottom, or they could run in opposite directions with respect to placement of the C- and N-termini. These structures are referred to as parallel and antiparallel beta sheets, respectively; both forms are found in proteins.

A third common secondary structure element in globular proteins is the beta turn. Beta turns are short segments (ca. five residues) of the amino acid sequence of the protein that allow the contiguous polypeptide to change direction. Because of the steric constraints imposed by such structures, residues making up turns most often contain small side chain groups. For this reason, glycine is commonly found associated with turns in globular proteins (Creighton, 1984).

Other regular secondary structures are sometimes found in proteins, such as the 3_{10} helix, poly-proline helices, and poly-glycine helices. These are, however, not common in globular proteins, and in many cases can be ignored (see Creighton (1984) for a discussion of these structures).

These regions of regular secondary structure are interspersed with sections of non-repeating unordered structure, commonly referred to as random coil. This does not imply that these regions of a protein are devoid of structure, but rather that the structure seen here is not repeating, or regular, nor does it fit into a specific category of secondary structure type. These regions of the protein are usually in greater dynamic flux than are the regions of regular secondary structure, and can thus play an important role in providing structural flexibility to the protein.

TERTIARY STRUCTURE

The next level in the hierarchy of protein structure is tertiary structure (figure 1.8). This term refers to how the elements of secondary structure

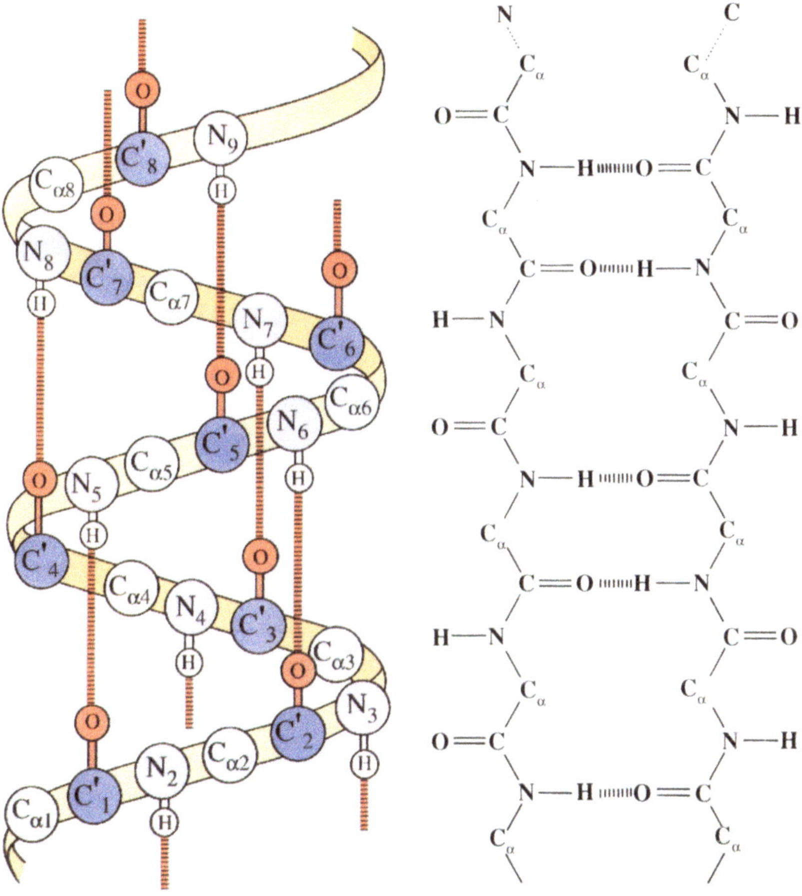

Figure 1.7 The right-handed alpha helix (left) and beta pleated sheet (right) secondary structures that are commonly found in globular proteins.

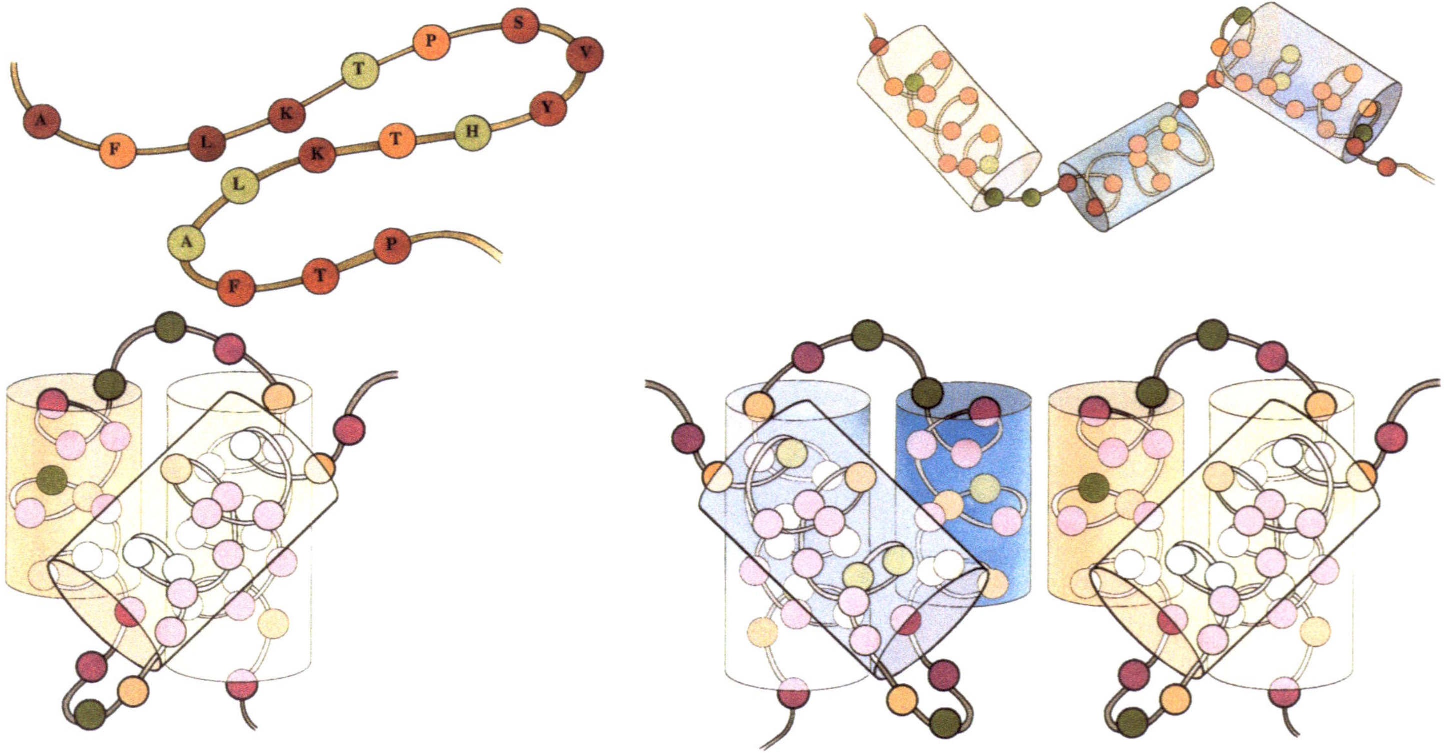

Figure 1.8 The hierarchy of protein structures: (1) primary structure or amino acid sequence, (2) secondary structure, (3) tertiary structure, and (4) quarternary structure.

arrange themselves in three dimensions in the folded protein, and how amino acid side chains interact with one another to form hydrogen bonds, electrostatic salt bridges, and hydrophobic-hydrophobic interactions (Branden and Tooze, 1991). Proteins fold into their native (i.e., naturally occurring biologically active) tertiary structure because of a variety of thermodynamic factors. Among these, one of the strongest driving forces is the need to shield nonpolar amino acids from the aqueous solvent into what has come to be known as the hydrophobic core of a protein. Likewise, folding favors interactions between polar amino acids and solvent molecules at the hydrophilic protein surface. Because of these types of forces, proteins often spontaneously fold into their native structures when presented with favorable solution conditions (Tanford, 1980).

Among the different side chain interactions that can occur in proteins is covalent bond formation through the oxidation of two sulfhydryl groups on cysteine residues. Such sulfur-sulfur bonds are known as disulfide bonds, and their presence in a folded protein provides an additional measure of stability to the native conformation. In multicysteine containing proteins, more than one arrangement of disulfide bonds may be possible. In such cases what almost always occurs is that only one specific set of disulfide bonds leads to an active protein. This is illustrated in figure 1.9 for the kringle 2 domain of human tissue plasminogen activator. Only the unique set of three disulfide bonds shown in figure 1.9 will lead to the correctly folded state of this protein (Vlahos et al., 1991).

The importance of protein tertiary structure cannot be overstated. It is the tertiary structure that gives a protein its overall shape and dimensions, and also provides a means of bringing into spacial proximity amino acid residues that may be distant in the linear sequence of the protein, but that need to come together to form the catalytic site of an enzyme, the binding pocket of a receptor, or a recognition site for the action of another protein. Thus tertiary structure is important for the following reasons:

- occlusion of hydrophobic residues from the polar solvent
- presentation of charged residues to solvent
- bringing groups that are distant along the sequence into close proximity for interactions
- providing the overall shape to the protein—important for establishing binding sites, sites for macromolecular recognition, etc

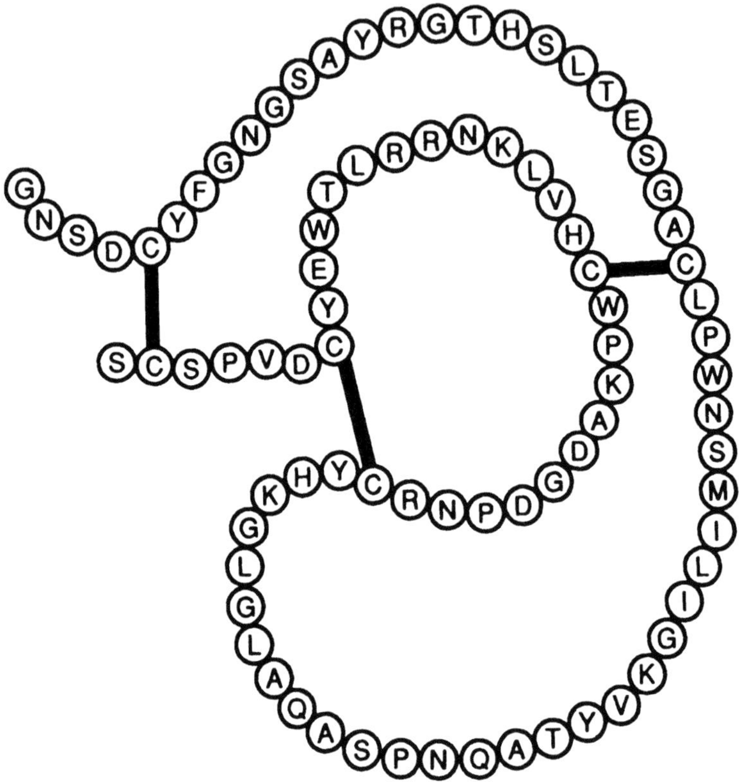

Figure 1.9 The amino acid sequence of the kringle 2 domain of tissue plasminogen activator showing the location of disulfide bonds between cysteine residues of the protein. Data taken from Vlahos et al. (1991).

QUATERNARY STRUCTURE

Often the active form of a protein is really a complex of several polypeptides that are held together by non-covalent (i.e., hydrogen bonds, electrostatic forces, and hydrophobic-hydrophobic interactions) and/or covalent (disulfide bonds) interactions. The individual polypeptides in such a complex are referred to as subunits of the protein. The arrangement of these subunits relative to one another defines the quaternary structure of the protein. The importance of protein quarternary structure is illustrated by the action of the blood protein hemoglobin. Hemoglobin is a heterotetramer composed of two alpha and two beta subunits. Each of these subunits contains a heme group with a central iron atom that forms the binding center for molecular oxygen. The four molecules of oxygen that bind to each hemoglobin molecule do not bind independently, but rather in a cooperative fashion. That is, the binding of oxygen to one heme in the protein enhances the binding affinity of the other three heme groups for oxygen. The binding and release of oxygen by hemoglobin in the lungs and muscles, respectively, is controlled by subtle conformational transitions within each subunit that change the relative orientation of the subunits with respect to one another. It is this change in quarternary structure that controls the cooperative binding of oxygen to hemoglobin, which is critical for life in higher organisms.

This hierarchy of protein structure is summarized in figure 1.8. As we shall see, much of the effort expended on protein analysis is aimed at assessing these different levels of protein structure.

MEMBRANE PROTEINS—A SPECIAL CASE

The structural motifs that we have discussed thus far pertain to all proteins. The majority of proteins that have been studied in detail have been water soluble components of cells. However, there is a large class of proteins that are not free floating in the cells, but rather are inserted into the phospholipid bilayers that make up biological membranes. These integral membrane proteins have some features that are distinct from the water soluble proteins, and need to be discussed here (Stryer, 1988).

The phospholipids of a biological membrane form a bilayer sheet with the highly charged phosphate head groups exposed to the two aqueous phases that are separated by the membrane, and the hydrophobic tails of the lipids forming an oily interior structure. To become associated with such a structure, membrane proteins must traverse the bilayer, and thus require long stretches of nonpolar amino acids for this purpose. When a section of polypeptide is removed from a protic solvent like water, it seeks to form intramolecular hydrogen bonds to compensate for the loss of hydrogen bonds that would have occurred with the solvent. There is, then, a tendency for peptides to form alpha helical structures in aprotic solvents, including the center of a phospholipid bilayer. Thus, most transmembrane proteins are believed to traverse the bilayer as alpha helices. Since a typical biological membrane has a hydrocarbon core cross-section of about 30 Å, and as we have just seen, each residue of an alpha helix represents a translation of 1.5 Å, it follows that about 20 residues are required to traverse the membrane as an alpha helix.

Can we predict which sections of a protein's amino acid sequence are likely to form such transmembrane alpha helices? The answer is yes, if we consider the thermodynamics associated with partitioning the various amino acid side chains between a polar and nonpolar solvent (Tanford, 1980). After first neutralizing the amino and carboxyl groups so that the partitioning properties corresponded more closely to that of an amino acid side chain within a protein, several workers measured the free energy of transfer for the 20 amino acids from water into various nonpolar solvents. While the free energies of transfer $\Delta G_{transfer}$ vary from solvent to solvent, certain trends can be discerned from these data. Not unexpectedly, highly charged residues, like lysine and glutamate, strongly prefer the polar aqueous solvent, while nonpolar residues, such as phenylalanine, have a strong propensity for partitioning into the nonpolar solvents. On the basis of this type of data, scales of hydrophobicity have been derived for the 20 amino acids within proteins (Kyte and Doolittle, 1982). Table 1.2 summarizes the most popular of these scaling systems, that described by Kyte and Doolittle (1982). By plotting the scale value as a function of amino acid number along the sequence of a protein, one

Table 1.2 Hydropathy indices for the naturally occurring amino acids based on the scale of Kyte and Doolittle (1982).

Amino Acid	Hydropathy Value
Ile	+4.5
Val	+4.2
Leu	+3.8
Phe	+2.8
Cys	+2.5
Met	+1.9
Ala	+1.8
Gly	−0.4
Thr	−0.7
Ser	−0.8
Trp	−0.9
Tyr	−1.3
Pro	−1.6
His	−3.2
Glu	−3.5
Asp	−3.5
Gln	−3.5
Asn	−3.5
Lys	−3.9
Arg	−4.5

obtains what is known as a hydropathy plot. These plots, as illustrated in figure 1.10, can be used to search for stretches of 20 or more amino acids, with hydropathy greater than some predetermined cutoff point, that are likely to form membrane-spanning helices. In practice, what is typically done is to use a floating window of seven residues to smooth out the hydropathy indexes, and thus gain a better average picture of where the most likely transmembrane regions may be. This method has proved surprisingly accurate in predicting the regions of transmembrane alpha helices in the few cases where subsequent biophysical methods have determined the actual disposition of the protein in the membrane (see, for example, Henderson et al., 1990).

SUMMARY

In this chapter we have seen how the side chain structures of the 20 amino acids provide a considerable range of chemical reactivities to

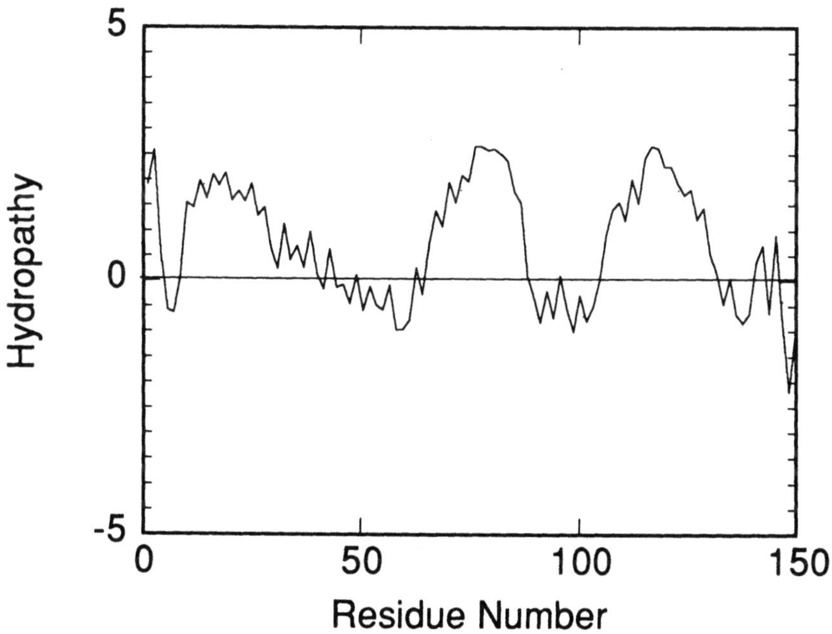

Figure 1.10 Hydropathy plot of the first 150 amino acids of subunit II of the cytochrome *c* oxidase of *Paracoccus denitrificans*. The plot predicts two regions of the polypeptide that traverse the membrane as alpha helices. The solid line represents the results of smoothing the hydropathy indices (Kyte and Doolittle, 1982) over a seven residue window.

proteins. Two or more amino acids can be joined together to form a peptide bond through a simple condensation reaction. We have seen that long strings of these amino acids can form polypeptides and proteins that are distinguished structurally on several levels: primary, secondary, tertiary, and quaternary. The information presented in this chapter provides a very cursory introduction to the topic of protein structure. Our aim here was to merely introduce the salient points that will be germane in our subsequent discussions of analytical methods for assessing protein structure and reactivity. For the reader who wishes to delve more deeply into the subject of protein structure, there are several excellent texts devoted to this subject that are listed among the references for this chapter.

References

Branden, C., and Tooze, J. (1991) *Introduction to Protein Structure,* Garland, New York.

Copeland, R. A. (1992) *Today's Chemist at Work,* **1,** 32–37.

Creighton, T. E. (1984) *Proteins, Structure and Molecular Properties,* W. H. Freeman, New York.

Dawson, R. M. C.; Elliott, D. C.; and Jones, K. M. (1969) *Data for Biochemical Research,* 2d. Ed., Oxford University Press, Oxford.

Henderson, R.; Baldwin, J. M.; Ceska, T. A.; Zemlin, F.; Beckmann, E.; and Downing, K. H. (1990) *J. Mol. Biol.,* **213,** 899–929.

Kyte, J., and Doolittle, R. F. (1982) *J. Mol. Biol.,* **157,** 105–132.

Pauling, L. (1960) *The Nature of the Chemical Bond,* 3d Ed., Cornell University Press, Ithaca, New York.

Ramachandran, G. N., and Sasisekharan, V. (1968) *Adv. Protein Chem.,* **23,** 283–437.

Stryer, L. (1988) *Biochemistry,* W. H. Freeman, New York.

Tanford, C. (1980) *The Hydrophobic Effect,* Wiley-Interscience, New York.

Vlahos, C. J.; Wilheim, O. G.; Hassell, T.; Jaskunas, S. R.; and Bang, N. U. (1991) *J. Biol. Chem.,* **266,** 10070–10072.

2

General Methods for Handling Proteins

In this chapter we shall review some general methods for the day-to-day handling of proteins, and some of the precautions that must be taken to ensure proper activity and stability of these biopolymers. Some useful methods for altering the solution conditions for protein samples will also be discussed here.

ROUTINE HANDLING OF PROTEINS

As will be discussed in Chapter 10, the native conformation of most proteins is only marginally stabilized over the unfolded or denatured state (for a review on protein stability see Pace, 1975). For this reason it is important that every precaution be taken to ensure maximal stability and activity of one's samples. Solution conditions and storage conditions can have a dramatic effect on these properties of the sample. Also, one sometimes finds that the conditions that lead to maximal activity of a protein are not the same as those conditions that lead to maximal stability. In these cases one must reach the best compromise between these conditions, on an individual empirical basis.

Proteins are in general more stable at lower temperatures. For routine handling, one should keep most proteins at ice temperature, or in a 4° C refrigerator. The length of time that a protein will be stable under these conditions will vary from protein to protein, but in general proteins can be stored for a few days at 4° C without significant denaturation. If an assay must be performed at room temperature or higher, the protein sample should be maintained at ice temperature until shortly before the assay is to take place (i.e., only as much time as is needed for temperature equilibration). Most proteins are stable at 25° C for at least an hour, so this does not present a major encumbrance to performing room temperature studies.

Aside from temperature, other solution conditions affect the stability of proteins. Solution pH can have a dramatic effect on protein stability. Again, there are variations in optimal pH from protein to protein, but most proteins are stabilized in the vicinity of physiological pH, i.e., pH 7.4, and this is usually a good starting point for optimization studies. For some proteins, especially enzymes, acid-base equilibria may play an important role in the activity of the protein; the optimal pH for activity can thus be very different from the optimal pH for stability (Fersht, 1985). For example, the mitochondrial enzyme cytochrome *c* oxidase displays maximal activity at around pH 6.0, while it is maintained most stably between pH 7.2 and 7.8.

Buffers for protein solutions should meet the same criteria as for buffers in general. Most importantly, the buffer system chosen should have good buffering capacity in the pH range that you wish to work in (i.e., the pK_a of the buffer should match the experimental pH range). Calbiochem-Behring provides a useful manual for biological buffers that explains the principles behind buffering, and lists the chemical properties of many common buffers (Doc No. CB0052-289). Some of the most commonly used buffers for protein work are listed in table 2.1, along with their pK_a values at 25° C. It should be kept in mind that the pK_a values for these buffers vary somewhat with temperature. Under most conditions, this is not a major concern, as the ΔpK_a/° C values are relatively small. These and other buffers can be purchased from a variety of biochemical suppliers (e.g., Sigma).

Salt bridges and other electrostatic interactions within a protein, between a protein and its substrate, or between two interacting proteins are greatly affected by ionic strength. At high ionic strengths these interactions are disrupted, leading to protein unfolding, or more commonly, decreased biological activity. Too low an ionic strength can, in some cases, lead to electrostatic interactions between protein molecules which are non-productive and decrease activity. Here again, the best starting point is usually that found in nature. Physiological ionic strength is about 0.15 M, and most proteins are stable under these conditions. The choice of salts to be used in maintaining ionic strength should take into consideration any special ion requirements of the protein. KCl is generally useful for this purpose, but if you know that your protein requires magnesium

Table 2.1 The pK_a values of some buffers that are commonly used in protein science.

Common Name	Chemical Name	Molecular Weight	pK_a
MES	2-(N-morpholino) ethanesulfonic acid	195.2	6.15
PIPES	piperazine-N,N′-bis(2-ethanesulfonic acid), monosodium salt	324.3	6.80
Imidazole	Imidazole	68.1	7.00
MOPS	3-(N-morpholino)propane sulfonic acid, sodium salt	231.2	7.20
TES	2-{[*tris*-(hydroxymethyl)methayl]amino} ethanesulfonic acid, sodium salt	251.2	7.50
HEPES	N-2-hydroxyethylpiperazine-N′-2-ethanesulfonic acid, sodium salt	260.3	7.55
HEPPS	N-2-hydroxyethylpiperazine-N′-3-propanesulfonic acid	252.3	8.00
Tricine	N-[(tris-hydroxymethyl)methyl]glycine	179.2	8.15
TRIS	tris-(hydroxymethyl)-aminomethane	121.1	8.30
CHES	2-(cyclohexylamino)ethanesulfonic acid	207.3	9.50
CAPS	3-(cyclohexylamino)propanesulfonic acid	221.3	10.40

for activity, you may wish to maintain ionic strength with $MgCl_2$, for example.

MEMBRANE PROTEINS

To maintain solubility, solutions of integral membrane proteins must be supplemented with detergents. Detergent micelles provide a mimic of the partitioning between polar and nonpolar surfaces that is provided by biological membranes. Detergents come in a variety of chemical forms, some nonionic, some anionic, some cationic, and some zwitterionic. The choice of which detergent to use for a specific protein depends on a number of factors. The most important factors to consider here are the ability of the detergent to solubilize the protein, and the effects of that detergent on protein activity. For some proteins, switching from one detergent to another can change the specific activity by as much as three orders of magnitude! Thus, it is worth a little experimentation to find the best detergent for your particular protein (see Hjelmeland and Chrambach, 1984, for a review of detergent solubilization of membrane proteins).

Regardless of what detergent one ultimately uses, the concentration of detergent is also an important consideration. To be effective in solubilizing membrane proteins, detergents must be capable of forming micelles, i.e., multimeric detergent assemblies with polar exteriors and an oily interior. In solution, detergent molecules are always in equilibrium between a monomeric state and the micelle form. Higher concentrations favor the micelle, and for every detergent there is a concentration above which all molecules added to solution will be incorporated into micelles. This concentration is referred to as the critical micellization concentration (CMC), and is dependent on the molecular properties of the detergent, the solution temperature, pH, and ionic strength. To effectively solubilize a protein, the detergent must be present in solution at a concentration greater than its CMC. On the other hand, too high a concentration of detergent can lead to deactivation of some proteins. Finding the optimum concentration of detergent for a specific protein is a matter of experimentation. Hjelmeland and Chrambach (1984) suggest that one experiment within a detergent to protein mass ratio range of 10:1 to 0.1:1. An excellent handbook on the properties of detergents and their use in solubilizing proteins is available from Calbiochem-Behring (Doc. No. 8183-10-89).

CONTAINERS FOR PROTEINS

Proteins adhere strongly to glass surfaces, so whenever possible glass should be avoided. Plastic containers, while still adsorbing some protein, ameliorate the problem significantly. Test tubes, pipette tips, transfer pipettes, microcentrifuge tubes, and cryotubes are available in low-protein-binding plastics from numerous suppliers. These should be used for containing and transferring proteins whenever possible.

For some applications one must use a glass container to temporarily hold a protein sample. When this is necessary, the degree of protein adsorption can be minimized by silanizing the glass surfaces to render them more hydrophobic. Seed (1989) has outlined the following protocol for silanizing laboratory glassware for protein work:

MATERIALS

1. Chlorotrimethylsilane or dichlorodimethylsilane (Dow Corning).
2. A vacuum desiccator with stopcock
3. A vacuum pump.

PROCEDURE

1. Place the desiccator in a working fume hood and place the glassware to be silanized into the desiccator.
2. Add 3 ml of chlorotrimethylsilane or dichlorodimethylsilane to the glassware.
3. Connect the desiccator to the vacuum pump and apply vacuum until the silane begins to boil. At this point close the stopcock and allow the evacuated and closed desiccator to stand until all of the liquid silane is gone (ca. 1–3 hours).
4. Open the desiccator in the hood, and leave it open for a few minutes to eliminate silane vapors.
5. Rinse glassware with distilled water before drying or autoclaving and storage.

NOTES

1. These silane liquids are flammable, and give off highly toxic vapors. Due caution should be exercised during this procedure.
2. Do not leave the vacuum pump connected to the desiccator with the stopcock open after the silane begins to boil. Doing so will evacuate the silane from the desiccator and thus minimize the effectiveness of the treatment.

STORAGE OF PROTEINS

For long term storage (more than a few days), proteins are best kept at very low temperatures, either in a −80° C freezer, or under liquid nitrogen (−196° C). Because proteins themselves have some antifreeze and buffering capacities, they are best stored at high concentrations. Proteins should be frozen and thawed rapidly, as this reduces the chances of sample degradation. For rapid freezing, samples should be immersed in a dry ice/ethanol bath until solid. Rapid thawing can be accomplished by placing the sample in a 37° C water bath until only a small piece of frozen material remains. At this point, the sample should be removed from the bath, mixed gently, and placed in ice to further thaw. Repeated freeze-thaw cycling is very destructive to proteins, and leads rapidly to denaturation. To avoid this, it is best to store proteins in highly concentrated, small volume aliquots. This way, only the amount of material needed for that day's work needs to be thawed, while the rest of the protein remains frozen. For example, in my laboratory we typically work with enzyme samples in the range of 10 μM. We store our enzyme at −80° C in 50 μl aliquots at a concentration of 200 μM or greater. At the start of the day, we remove one aliquot of enzyme, and dilute it with 950 μl of buffer to obtain 1.0 ml of a 10 μM solution.

If proteins are stored in −20° C freezers, addition of 50% glycerol or 0.25 M sucrose will enhance their long term stability. This is particularly important for "frost free" freezers that cycle through defrost cycles.

An important factor that limits the shelf life of protein solutions is bacterial contamination. It is therefore important that every precaution be taken to reduce the risks of such contamination. Cryovials or other storage containers for proteins should be sterilized before use. Any buffers or other reagents that will be used to dilute the protein sample before storage should be sterilized either by autoclaving or by filtration through a sterile 0.22 micron (μ) filter. Of course, protein solutions themselves cannot be autoclaved, since this would lead to thermal denaturation of the protein (see Chapter 10). However, the protein solution should be passed through a sterile 0.22 μ filter just prior to storage. Filtration can lead to some protein loss, due to adsorption onto the filter material, but modern filter units are made of low protein binding materials that reduce this problem. An additional guard against bacterial contamination is to add 0.02% sodium azide to the protein solution as an anti-bacterial agent (Scopes, 1982). This should only be used, however, when one has predetermined that azide does not affect the activity of the protein.

Proteins that contain free sulfydryl groups can, over time, aggregate through the formation of intermolecular disulfide bonds. This process is an oxidative one, and is thus accelerated by oxygen in the air, and by metal ions. Purging containers with nitrogen or argon gas just before storage can help to retard intermolecular disulfide formation by limiting the oxygen present in the sample (argon has the advantage of being denser than air so that it does no escape as readily from the container). Sparging buffers with these gases, prior to their use with protein samples also helps in this regard. However, sparging solutions that already contain proteins is not recommended, since bubbling tends to denature proteins.

Addition of 1 mM EDTA to protein samples will chelate metal ions and thus help to retard disulfide bond formation. EDTA is also useful in inhibiting the action of metalloproteases (see below). If one's protein contains only free cysteine residues (not intramolecular disulfides), then reducing agents, such as dithiothreitol (DTT) can be added to the sample at millimolar concentrations to prevent disulfide bond formation. If intramolecular disulfide bonds are present in one's protein, reducing agents can still be used to retard aggregate formation, but lower concentrations are recommended (i.e., micromolar). Some experimentation is required to find the best set of conditions for each individual protein.

If a protein has been purified to homogeneity (i.e., it is the only biopolymer present in the sample), then problems of proteolytic digestion should be minimal. Proteases are enzymes that act to cleave peptide bonds in other proteins, and are present in most cells. Thus, early in the purification scheme of a protein, there will usually be proteases present, and these must be inhibited to prevent degradation of the target protein. If one is working with samples of impure protein or cell lysates, it is a good idea to include protease inhibitors in the working buffers. Table 2.2 lists some of the common types of proteases that are encountered in biochemical work, and some chemical-based inhibitors of these enzymes. For general protection of protein samples from proteolysis, it is a common practice to combine several of these inhibitors into a protease inhibitor "cocktail," and add this to the sample. Boehringer Mannheim recommends the following cocktail for general applications: leupeptin, 0.5 mg/liter; EDTA, 1 mM (3g/liter); pepstatin, 0.7 mg/liter; and PMSF, 0.2 mM (35 mg/liter).

Protease inhibitors are available from a large number of sources, in-

Table 2.2 Some common proteases encountered in biological samples and chemical based inhibitors of these enzymes. Adapted from the Boehringer Mannheim bulletin "Protease Inhibitors."

Protease Type	Inhibitor
Papain, Trypsin, Cathespsin B	Antipain-dihydrochloride
Serine proteases	APMSF (amidino-phenylmethane sulfonyl fluoride)
	PMSF (phenylmethane sulfonyl fluoride)
Trypsin, Plasmin, Chymotrypsin, Kallikrein	Aprotinin
Aminopeptidases	Bestatin
α-, β-, γ-, δ-Chymotrypsin	Chymostatin
Thiol proteases	E-64 (Boehringer Mannheim), or PMSF (phenylmethane sulfonyl fluoride)
Metalloproteases	EDTA
Serine and Thiol proteases	Leupeptin
Acid proteases	Pepstatin
Metallo Endopeptidases (for example, Thermolysis)	Phosphoramidon

cluding Sigma, Calbiochem., and Boehringer Mannheim. Even when a protein sample has been purified and is free of proteases, one must take precautions not to inadvertently introduce proteases into the sample. A common source of these proteases is the investigators themselves. For this reason, one should always take the precaution of wearing gloves when handling protein samples, and avoid standing directly over an open tube of sample in such a way that hair, particles of skin, etc., might fall into the sample.

REMOVAL OF SMALL MOLECULES FROM PROTEIN SOLUTIONS

Many times proteins are present in solution along with small molecular weight components that need to be removed prior to a particular analytical procedure or storage. For example, if the final step in the purification of a protein is to elute it from an ion exchange column with a salt gradient, the final protein sample may have a higher than optimal concentration of salt present. Methods for removing small molecular weight species from protein solution all take advantage of the large molecular weight difference between proteins and polypeptides, and small molecules. Because salt removal is one of the most common motivations in using these methods, the process is sometimes referred to as "desalting" proteins.

Dialysis

Dialysis is one of the oldest methods for removing selectively small molecular weight components from solution. The method is based on the availability of cellulose membranes that have well-defined molecular weight cutoffs (pore sizes) that allow small molecules to traverse the membrane, while restricting the movement of larger molecules (Craig and King, 1957; Cooper, 1977). The molecular weight cutoff is defined as "that solute molecular weight that is retained to an extent of 90%." These membranes are supplied commercially as rolls of tubing in varying diameters and molecular weight cutoffs (available from Spectrum, Houston Texas). The larger the molecular weight cutoff, the more rapidly small solutes will permeate the membrane. One must bear in mind however, that these molecular weight cutoffs represent a nominal value, not a sharp cutoff. Thus, one should use a larger molecular weight cutoff for rapid equilibration, but the cutoff must also be well below the molecular weight of the protein that one wishes to retain.

To use dialysis tubing, one must first hydrate it to give flexibility, and then seal it on one end. The protein sample is then placed within the tubing, and the other end of the tube is sealed. The tube is then placed in a solution of buffer that has the final molecular composition desired for the protein. With time, molecules that are small enough to pass through the pores of the dialysis tubing will equilibrate between the internal and external solutions.

PREPARATION OF DIALYSIS TUBING

1. Cut a section of tubing long enough to accommodate the volume of protein to be dialyzed, and to allow room on either end for sealing.
2. Place the tubing in a beaker containing 100 mM $NaHCO_3$ and 10 mM EDTA, pH 7.0.
3. Place the beaker on a hot plate and bring the solution to a boil. Boil tubing for 5 min.
4. Decant off the solution and replace with distilled water. Soak the dialysis tubing for 10 min. Repeat the distilled water rinse a minimum of four times.
5. Use the tubing immediately. Alternatively, a large amount of tubing can be prepared and stored for future use. Store dialysis tubing at 4° C in 20% ethanol (aqueous) to prevent bacterial growth. Before use, wash the tubing with copious amounts of distilled water.

Note: Prepared dialysis tubing should always be handled with gloved hands to prevent accidental contamination with proteases.

USE OF DIALYSIS TUBING

1. Seal one end of a prepared segment of tubing by either tying a double knot in the tube, or by using a dialysis bag closure (available from Spectrum).
2. Fill the tube with distilled water or the sample buffer. Gently squeeze the tube from the open end and check to make sure that the sealed end is not leaking.
3. Empty the tubing and refill with the protein sample. Seal the open end of the tubing as above. Try to seal the tube close to the protein solution, but leaving a small head space for expansion. Having a large space between the solution and the closure will result in dilution of the sample, as buffer fills the space during dialysis.

4. Fill a beaker with at least 200 volumes of the buffer (at 4° C) that you wish to dialyze the sample against. Place a stir bar and the sealed dialysis bag in the beaker, and cover the beaker.
5. Place the beaker on a stir plate at 4° C, and start gently stirring the solution (do not stir too rapidly, as this could cause rupture of the dialysis bag). Dialyze the sample for 12 to 24 hours, replacing the dialysis buffer at least three times during this interval.

The above procedure works well for protein solutions from 1 to tens of ml. For very small volumes (less than 1 ml), special apparatus are available, such as the Spetra/Por Dispodialyzer (from Spectrum), that allow reduced volume dialysis. Alternatively, a simple device for dialyzing small volumes can be constructed by boring a hole in the cap of a microcentrifuge tube (Overall, 1987; Falson, 1992). The sample is placed in the tube, and a sheet of dialysis tubing is placed over the orifice of the tube. The cap is then carefully closed so that the dialysis tubing forms a barrier between the solution inside the tube and the outside environment. The tube is then placed upside down in the bottom of a floating microcentrifuge holder (available from Ann Arbor Plastics) within a beaker of dialysis buffer or it can be taped to the side of the beaker (see figure 2.1). Of course, the success of this method depends critically on forming a leakproof seal with the dialysis tubing. Volumes as small as 10 μl can be effectively dialyzed in this way.

Gel Filtration Chromatography

An alternative method for removing small molecular weight species from protein solutions is the use of gel filtration chromatography (Cooper, 1977; Scopes, 1982). Because of the packing geometry of the stationary phase material, small molecules can enter the interstitial space between gel particles where their migration through the column is retarded. The degree of retardation will be dependent on the resident time of the molecule in the interstitial space, which in turn is inversely related to the size of the molecule. If a molecule is too large to enter the interstitial space, then its migration will not be retarded by the stationary phase, and it will elute at the void volume of the column. The size of the interstitial spaces, and thus the minimum molecular weight of the molecules that will not be retarded on the column (referred to as the exclusion limit of the column), varies depending on the type of stationary phase material used. Tables of commercially available stationary phase materi-

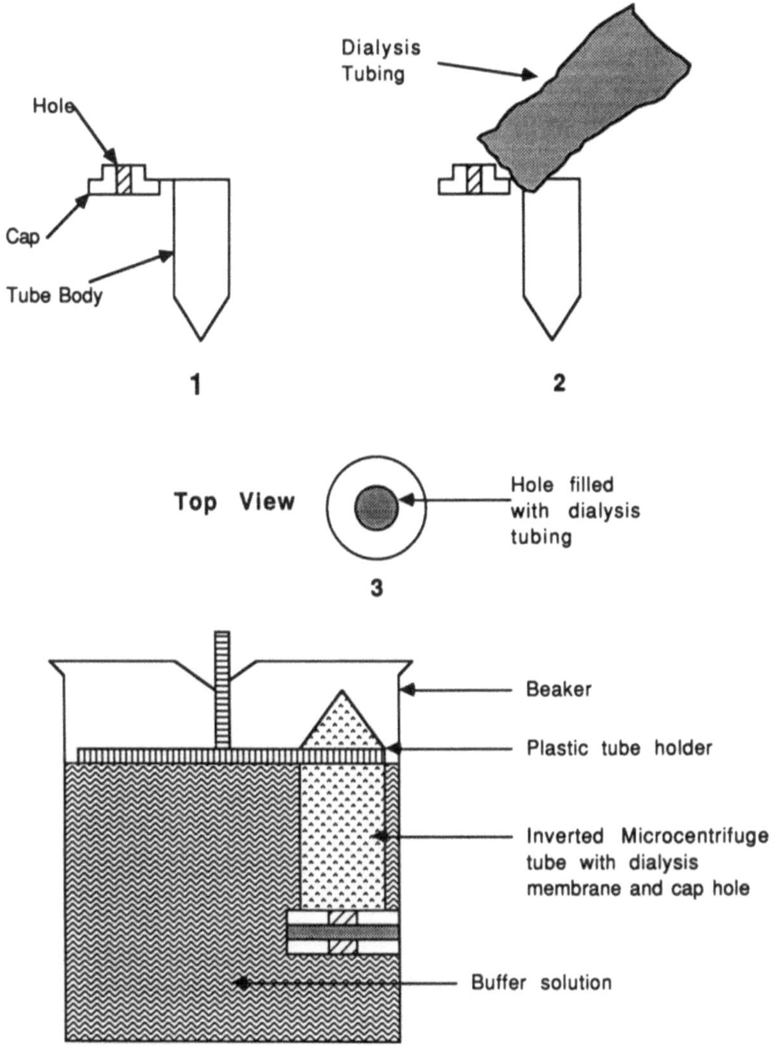

Figure 2.1 Apparatus for microdialysis of small volumes of protein solution. (1) A hole is formed in the cap of a microcentrifuge tube by pushing the end of a hot Pasteur pipette through the plastic. (2) After adding sample to the bottom of the tube, a sheet of dialysis membrane is laid over the top of the tube and the cap is carefully sealed over the membrane. Membrane extending beyond the cap is trimmed with a pair of scissors. A top view of the assembled apparatus is shown in (3). (4) The apparatus is placed inverted into a beaker of buffer solution for dialysis. The tube is held in place by a plastic microtube holder, or can be taped to the side of the beaker with laboratory tape (Overall, 1987).

als for gel filtration, comparing their useful fractionation ranges and exclusion limits have been published (Cooper, 1977) and are available from the manufacturers. For proteins of molecular weights less than the exclusion limit, gel filtration chromatography can be successfully used to separate individual proteins from a mixture. This method is thus commonly used as part of the purification schemes for many proteins. The use of gel filtration chromatography for protein purification has been extensively covered in several texts, and is also discussed in Chapter 4. Pharmacia also provides an excellent booklet on the theory and practice of gel filtration chromatography (*Gel Filtration, Theory and Practice,* available from Pharmacia Fine Chemicals). Here, we will focus exclusively on its use for desalting protein samples.

As an illustrative example, let us say that we wish to change the salt concentration from 500 mM NaCl to 150 mM KCl for a sample of a protein of molecular weight 30,000 Da. We might choose to use a Bio-Gel P10 (Bio-Rad) or Sephadex G-15 (Pharmacia) column for this purpose. The first steps in preparing such a column would be to rehydrate the dry gel material, and pour the column. It is critical that one follow the manufacturer's detailed instructions for both of these steps. Often in the manufacture of gel materials, small particles, referred to as fines, are generated (Cooper, 1977). The inclusion of fines in the bed of a column can lead to irregularities in flow that reduce the effectiveness of the column. Fines can be removed during rehydration by suspending the gel material in a four-fold excess of equilibration buffer (i.e., the buffer that will be used to run the column and make up the final solution for the sample. In our example this would be the buffer containing 150 mM KCl). After gently mixing the suspension by inversion in a closed vessel, the suspension is placed in a graduated cylinder and the gel is allowed to settle by gravity. When about 90% of the gel has settled, the solution above the gel is removed by suction (do not decant the solution). This procedure should be repeated twice during the manufacturer's recommended time for rehydration (Cooper, 1977).

Rehydration and pouring of the column should always be done at the temperature at which one intends to perform the chromatography. Thermal expansion and contraction of the gel material can lead to channels in the bed material which will severely reduce the resolving power of the column. For similar reasons, it is important to remove gas bubbles from the rehydrated gel. This is best accomplished just prior to pouring the column by transferring the gel slurry to a filtration flask and applying a vacuum for about 20 min. During this time, the slurry can be gently swirled to enhance gas elimination.

The column itself should be made of plastic or treated glass, and must have a fritted bottom to allow solvent and solutes to pass while retaining the gel material. Columns for protein chromatography are available from Bio-Rad, Pharmacia, and other manufacturers. The column should be thoroughly cleaned, dried, and equilibrated at the running temperature prior to pouring the bed. The size of the column depends on the volume of protein that one wishes to desalt. For desalting purposes, the sample volume can be as large as 25% of the bed volume of the column. The volume of the column can be computed by the formula for the volume of a cylinder:

$$V = \pi r^2 h$$

where V is the volume (in ml), r is the radius of the column (in cm), and h is the height of the bed (also in cm). This formula should be used to select a column of appropriate size. To determine exactly the bed volume, one closes the stopcock or other closure device on the column bottom and fills the column with the desired volume of distilled water (measured in a graduated cylinder). The top of the water volume is then marked on the outside of the column with a piece of tape or with a waterproof marking pen, and the water is drained out. While the water is draining through the column, tap the walls gently near the bottom. This will help release any air bubbles that might be trapped between the frit and the bottom of the column. With the bottom of the column closed, a few ml of buffer are placed in the column to give a cushion between the fritted bottom and the gel bed that is about to be poured. Next one pours the column bed into the column using a glass stir rod against the side of the column to gently guide the slurry in. Care must be taken to avoid bubbles during this process. One should attempt to add all of the gel bed at one time to the column, approximately estimating the volume of slurry based on the dilutions used during rehydration. To accommodate this much slurry, one must use a column extender or funnel that is available from most of the column manufacturers. Allow the slurry to begin to settle, and then open the closure at the bottom of the column and allow the solvent to drain. Carefully monitor the column during this process to ensure that the column does not dry out. Continually add more buffer to the top of the column during settling, until all of the gel has settled. If it appears that one needs more gel material to fill the column to the mark, then more slurry should be added during settling. Do not allow the gel to settle and then add additional material on top of this, as this will cause a discontinuity at the bed interfaces that will reduce the resolving power of the column (Cooper, 1977). Once the gel has completely settled, close the stopcock at the bottom of the column and leave a few cm of buffer above the top of the column bed. To run the column, perform the following steps:

COLUMN RUNNING PROCEDURE

1. Equilibrate the column by running two or three column volumes of the running buffer (the 150 mM KCl in our example) through the column. When the meniscus of the buffer reaches the top of the column bed, close the bottom of the column.
2. Apply the protein sample to the column by touching the tip of a plastic transfer pipette, containing the sample, to the wall of the column just above the bed. Gently release the protein onto the column wall while rotating the pipette around the wall of the column. This should be done very carefully so as not to disturb the top of the bed. Once the sample is completely applied, open the stopcock and allow the sample to enter the column bed. Close the stopcock when the sample meniscus reaches the top of the bed.
3. Carefully apply a volume of buffer equal to that of the sample to the column, again being careful not to disturb the bed. Run this into the column bed as above.
4. Measure out two column volumes of buffer and add as much of this as will fit into the column. Open the stopcock and allow the buffer and sample to begin eluting from the column. Continue to apply buffer until it has all been added to the column.
5. While the column is running, collect fractions of about 1/20 the bed volume throughout the run. Close the stopcock when the buffer meniscus reaches the top of the column bed.
6. Analyze the fractions for protein by absorbance at 214 or 280 nm, or by one of the methods described in Chapter 3, and pool those fractions that contain protein.
7. Re-equilibrate the column with three bed volumes of buffer to remove residual salt before the next run. Store the column at the temperature at which it will be run. If the column is to be stored at room temperature, and will not be used frequently, it is a good idea to store it equilibrated with running buffer containing 0.02% azide to prevent bacterial contamination.

A very convenient alternative to preparing and storing a gel filtration column as above, is to use a prepackaged disposable desalting column. These are available from several sources. My own laboratory has found the Bio-Rad disposable P-10 columns to be very effective in this regard.

These come as prepackaged 10 ml bed volume disposable columns with a fritted adapter at the top of the bed that prevents bed disruption during sample application. The columns come equilibrated with azide-containing buffer, so one should avoid contact with the buffer (wear gloves). The following procedure has been suggested by Bio-Rad for the use of these columns.

1. Remove the top cap and decant the buffer from the column. Snap off the bottom seal, and clamp the column into position for running.
2. Equilibrate the column with 25 ml of running buffer. Allow the buffer to run through the column until no further dripping from the bottom occurs (the columns are designed so that they will not dry out under these conditions).
3. Apply 3.0 ml of sample to the column and allow this to run into the bed until dripping from the bottom ceases. If the sample is less than 3.0 ml, dilute it with buffer to achieve this volume.
4. Apply 4 ml of running buffer to the top of the column, and collect 0.5 ml fractions until the column ceases to drip. Pool the fractions containing protein, and discard the column.

These columns are extremely easy to use, and save the researcher considerable time, although they are more expensive than conventionally prepared columns. Disposable desalting columns are also very useful for applications requiring the removal of radiolabeled small molecules from protein solutions. They can be used and then disposed of with other solid radioactive waste, thus eliminating the need to keep a radioactive column in the laboratory.

For smaller volumes of protein, one can combine gel filtration chromatography with centrifugation through the use of spin columns, for rapid desalting. The following procedure, provided by Dr. Michael Tota of Merck Research Laboratories, works well.

1. Prepare Sephadex G-15 (or other suitable resin) by weighing out 50 g of the dry resin, and rehydrating this with 500 ml of distilled water. After settling, aspirate off 200 ml of water. The flask should now contain approximately equal volumes of swollen resin and water.
2. Swirl the resin gently to resuspend it, and pipette 2 ml of the suspension into a disposable spin column (Dowex or Bio-Rad). Let this gravity drip to pack the column.
3. Equilibrate the column twice with 2 ml of the running buffer. Let the buffer gravity drip through the column.

4. Place the column in a 95 × 16 mm plastic test tube. Centrifuge this for 5 minutes in a Beckman AccuSpin table top centrifuge (or equivalent) at 1,700 × g.
5. Transfer the column to a clean test tube, and add 100 μl (or less) of the sample. Centrifuge the column as above, and collect the desalted protein from the test tube.

METHODS FOR CONCENTRATING PROTEINS

As described above, proteins are best stored as concentrated solutions. However, in the course of purification, the concentration of a protein solution is often reduced, and one must find ways to increase the final concentration of the sample prior to storage. Additionally, there are times when one wishes to concentrate protein samples for application of an analytical method of limited sensitivity. Some methods for concentrating protein solutions are given below. It should be noted that some of these methods are denaturing to the sample, and should therefore only be used when the sample's structural integrity is not important.

TCA Precipitation

Most proteins will precipitate out of solution when exposed to high concentrations of trichloroacetic acid (TCA). This method is quick and easy to perform, but leads to denaturation of the protein that is not necessarily reversible. For this reason, TCA precipitation should only be used for analytical purposes that do not require the native conformation of the protein.

MATERIALS

1. Ice cold 6.1N trichloroacetic acid (TCA; available from Sigma, Aldrich, and other sources)
2. Ethanol
3. Microcentrifuge tubes
4. Microcentrifuge
5. Buffer.

PROCEDURE

1. Place 900 μl of the protein sample in a microcentrifuge tube (the sample should contain ≥ 5 μg of protein) and add 100 μl of TCA. Cap the tube and mix by inversion or vortexing.
2. Place the tube on ice and incubate for 30 min.
3. Centrifuge in a microcentrifuge for 5 min.
4. Aspirate off the supernatant quickly.
5. Wash the pellet twice with 100 μl of ethanol. Aspirate away all of the liquid.
6. Resuspend the pellet in the desired volume of buffer.

Ammonium Sulfate Precipitation

Under high salt conditions proteins tend to precipitate from solution. Ammonium sulfate has been found to be particularly useful in this regard. Often proteins that have been precipitated by ammonium sulfate can be resuspended in buffer solution with retention of their biological activity (although this is not universally true). For this reason, ammonium sulfate precipitation can also be used in protein purification. During purification one takes advantage of the fact that different proteins will precipitate at different concentrations (usually expressed in terms of percent saturation) of ammonium sulfate (Cooper, 1977; Scopes, 1982). For bulk protein precipitation, however, it is best to simply jump the ammonium sulfate concentration to a value where most proteins will precipitate. Bollag and Edelstein (1991) recommend using 85% saturation to achieve maximum precipitation of a broad range of proteins. A general procedure for precipitating proteins is described below.

1. Place the protein solution in a beaker with a stir bar. Place the beaker within a tub of ice, and place the tub on top of a magnetic stir plate.
2. Begin stirring the solution. Slowly sift in 61.2 g of solid ammonium sulfate per 100 ml of solution. It is best to add the ammonium sulfate to the vortex of the stirring solution and allow all of the solid to dissolve before the next addition. Adjust the pH of the solution with NaOH solution throughout this step.
3. Centrifuge the sample for 15 minutes at 15,000 × g. Decant off the supernatant.

4. Resuspend the protein pellet in the desired volume of buffer. Residual ammonium sulfate should be removed by one of the desalting methods described previously.

PEG Precipitation

Another relatively gentle method for precipitating proteins is to increase the concentration of polyethylene glycol (PEG) in solution above a critical concentration (Ingham, 1984). The vast majority of proteins will precipitate at 30% (w/v) PEG, and this is the concentration recommended for general protein precipitation. The procedure is similar to that used for ammonium sulfate precipitation. One adds 1.5 ml of a 50% (w/v) solution of PEG-6000 (available from Sigma) for each ml of protein solution, and begins to mix the solution gently. The solution is incubated, with mixing, for 1 hour and is then centrifuged (as per ammonium sulfate, above). The supernatant is discarded and the pellet is resuspended in the desired volume of buffer.

One problem with this method is the subsequent removal of residual PEG. Desalting columns are not much use for removal of PEG. The most straightforward means of removing residual PEG is by dialysis (described above).

Dialysis Methods for Concentrating Proteins

Earlier in this chapter we discussed dialysis as a means of desalting protein solutions. Dialysis can also be used to concentrate a protein solution if, instead of dialyzing against an aqueous buffer solution, one dialyzes against a hygroscopic material. The two materials most commonly used in this regard are polyethylene glycol (PEG) and sephadex G-200 (any of the sephadex resins could be used here, but sephadex G-200 adsorbs the highest amount of water per gram of dry resin; Cooper, 1977; Scopes, 1982). One follows the procedure described above to prepare and fill the dialysis tubing. Then the tubing is buried within a dry layer of PEG or Sephadex in a pyrex casserole dish. The dish is placed at 4° C and is tilted by propping it up on one end so that liquid will run off. When the bed material becomes moist, it is replaced with fresh material. This is a very gentle, although time consuming method, that is generally applicable to most proteins.

A second dialysis method for concentrating samples is vacuum dialysis. Here the protein sample within the dialysis bag is suspended in a filtration flask, and the flask is attached to a vacuum line, as illustrated in figure 2.2. The pressure differential across the membrane forces solvent (and small solutes) out of the bag, while the protein is retained. Over time the volume within the bag is reduced, and the protein concentration is thus increased. Commercial units for vacuum dialysis are available from Spectrum (Micro-ProDiCon).

Ultrafiltration

A number of devices are commercially available for concentrating protein solutions that make use of membranes of well-defined pore size that selectively pass solvent and solutes below a critical molecular weight. The protein solution is forced through the membrane by the application of pressure or centrifugal force. Solvent and small solutes pass through the membrane as a result of the applied force, but the larger molecular weight protein molecules are retained. Amicon Corporation and a num-

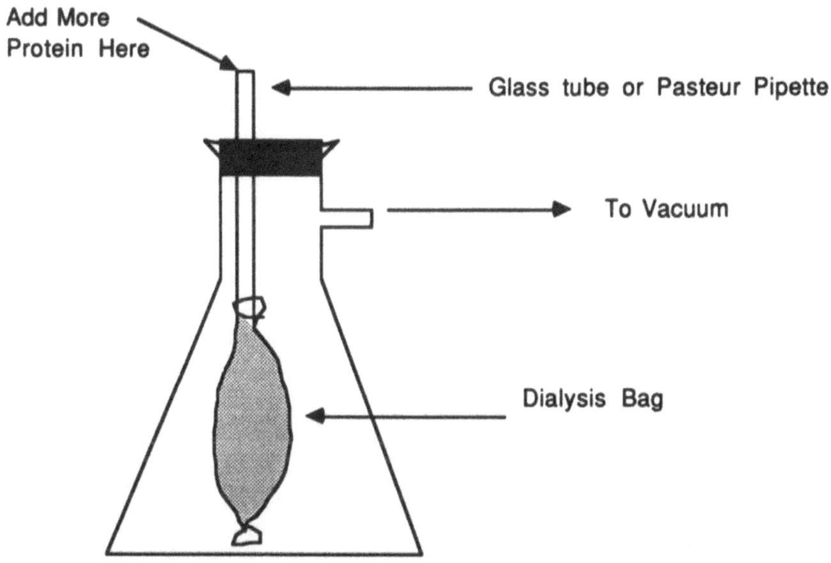

Figure 2.2 Apparatus for vacuum dialysis for concentrating protein solutions. The dialysis bag is attached to the end of a glass tube with a small rubber band. The other end of the dialysis bag is sealed with double knots or a commercial closure. Protein is continuously added to the dialysis bag through the top of the glass tube. When vacuum is applied, buffer and small molecular weight solutes permeate the dialysis tubing and collect in the bottom of the flask.

ber of other suppliers produce devices based on this technology for concentrating protein solutions from hundreds of ml down to less than 1 ml. Some of these devices are illustrated in figure 2.3. These systems are very effective in concentrating proteins of various sizes in a reasonably short amount of time (usually less than 2 hours). As with dialysis tubing, the membranes come in different molecular weight cutoffs, and one must choose a membrane that is appropriate for the size of one's target protein. Detailed descriptions of these products and their proper use are available from the manufacturers.

Lyophilization (Freeze Drying)

Some, but definitely not all, proteins can be stably stored as freeze dried powders, that can be subsequently rehydrated. Lyophilization, or freeze drying, is performed by freezing the protein solution as a thin film on the surface of a glass vessel (usually a test tube or round bottom flask that is immersed in a dry ice/ethanol bath), and then sublimating the solvent by application of a vacuum (Everse and Stolzenbach, 1971). Commercial units for freeze drying are available from a number of manufacturers, and are distributed by most of the large scientific supply houses. One caution that should be considered in lyophilization is that all nonvolatile solutes will be dried down by this method. Thus, if one's protein sample is in a buffered solution, the buffer and any salts present in solution will make up part of the lyophilized powder resulting from freeze drying. If this is undesirable, the sample should be dialyzed against distilled water, or a volatile buffer before lyophilization (Everse and Stolzenbach, 1971). A table of volatile buffers that are appropriate for this purpose has been compiled by Deutscher (1990). For buffering protein solutions in the physiological pH range, 100 mM ammonium bicarbonate (pH 7.9) works well.

The freeze drying process is accelerated by spreading the protein sample over as large a surface area as possible. This is best done by using a vessel of large volume relative to one's sample volume (for example, a 50 ml flask for a 1 ml sample). The walls of the flask can be coated with a thin film of solution by rapidly swirling the sample within a flask that is immersed in a dry ice/ethanol bath, until the solution is completely frozen. Once frozen, the sample flask should be transported, on dry ice, to the lyophilizing unit, and vacuum should be applied as soon as possible.

An alternative to lyophilization is vacuum centrifugation. Here the sample is placed in the bottom of a small tube and centrifuged while a

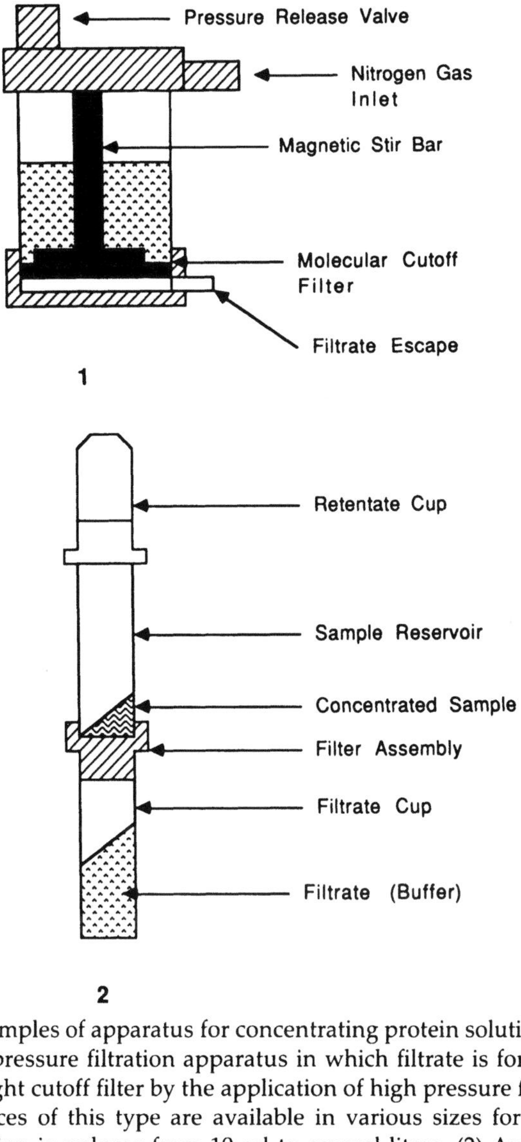

2

Figure 2.3 Examples of apparatus for concentrating protein solutions by ultrafiltration. (1) A pressure filtration apparatus in which filtrate is forced through a molecular weight cutoff filter by the application of high pressure from a gas (N_2) cylinder. Devices of this type are available in various sizes for concentrating solutions ranging in volume from 10 ml to several liters. (2) An apparatus for concentrating protein solutions by the application of centrifugal force. When placed in a centrifuge and spun, buffer and small molecular weight solutes (filtrate) pass through the molecular weight cutoff filter and collect in the filtrate cup. The protein sample (retentate) is concentrated and remains above the filter in the sample reservoir. After the sample is concentrated to the desired volume, the filtrate cup is removed, the apparatus is inverted and the retentate is collected in the retentate cup by centrifuging the apparatus again. These and other ultrafiltration devices are available from Amicon.

strong vacuum is applied. The centrifugal force keeps the sample within the tube during the process. In some cases the rate of solvent removal can be accelerated by heating the rotor to 45° C during centrifugation. Commercial units, such as the Savant Speed Vac Concentrator and Labconco CentriVap Concentrator, are available for this purpose.

It is important to recognize that lyophilization can be denaturing to some proteins. If one wishes to concentrate samples for analytical purposes that do not require native protein, then the methods described here can be used without hesitation. However, if one wishes to use lyophilization for long term storage of proteins, with retention of biological activity, pilot studies should be performed to determine the appropriateness of the method for the specific protein of interest.

References

Bollag, D. M., and Edelstein, S. J. (1991) *Protein Methods*, Wiley, New York.

Cooper, T. (1977) *The Tools of Biochemistry*, Wiley, New York.

Craig, L. C., and King, T. P. (1957) *J. Am. Chem. Soc.*, **79,** 3729.

Deutscher, M. P. (1990) *Guide to Protein Purification* (*Meth. Enzymol.*, **182**), Academic Press, San Diego.

Everse, J., and Stolzenbach, F. E. (1971) *Meth. Enzymol.*, **22,** 33–39.

Falson, P. (1992) *Biotechniques*, **13,** 20–21.

Fersht, A. (1985) *Enzyme Structure and Function*, 2d ed., W. H. Freeman, New York.

Hjelmeland, L. M., and Chrambach, A. (1984) *Meth. Enzymol.*, **104,** 305–318.

Ingham, K. C. (1984) *Meth. Enzymol.*, **104,** 351–355.

Overall, C. M. (1987) *Analyt. Biochem.*, **165,** 208–214.

Pace, C. N. (1975) *CRC Crit. Rev. Biochem.*, **3,** 1–43.

Scopes, R. K. (1982) *Protein Purification: Principles and Practice*, Springer-Verlag, New York.

Seed, B. (1989) in *Short Protocols in Molecular Biology* (Ausubel, F. M. et al., Ed.) Wiley, New York, p. 365.

3

Methods for Protein Quantitation

Numerous methods for determining the amount of protein present in a sample have been devised over the years. Gravimetric methods for lyophilized (freeze dried) proteins have been used, as well as methods based on elemental analysis, such as the Kjeldahl nitrogen assay (Kjeldahl, 1983). These methods have largely been replaced by colorimetric and spectroscopic methods which offer convenient and reasonably accurate estimations of protein content in solution. Since these are the most commonly used methods for protein quantitation, we shall discuss them in detail here. The reader who wishes to pursue other methods for protein concentration determinations is referred to the excellent review by Darbre (1986). A comparison of the most common methods for protein concentration analysis has recently been made by Bio-Rad Laboratories, and is summarized in table 3.1.

COLORIMETRIC METHODS

Colorimetric assays for proteins are based on the fact that certain metal ions and dyes bind to proteins in a specific mass ratio, and upon binding become intensely colored. Within a specific range of protein concentration, these reagents will give rise to an absorption band whose intensity is linearly proportional to the protein concentration of a solution in mass/volume units. The three most commonly used of these methods are: the Biuret assay, the Lowry assay, and the Bradford assay.

Table 3.1 Comparison of methods for protein concentration determination.

Method/Detection Limits	Chemical Interferences	Protein-Protein Variability	Speed and Difficulty of Method
Biuret/100μg	moderate	low	moderate/simple
Lowry/1μg	great	large	moderate/moderate
Bradford/1μg	low	large	rapid/simple
Absorbance at 280 nm/10μg	moderate	large	rapid/simple
Kjeldahl Nitrogen Assay/1μg	moderate	low	slow/complex

The Biruet Assay (Useful Detection Range: 0.5 to 80 mg/ml)

This assay is based on copper binding to proteins under alkaline conditions. The mode of interaction between the protein and copper is not well understood, but it was recognized early on that the intensity of the charge-transfer absorption band resulting from the copper-protein complex is linearly proportional to the mass of protein present in solution. This is a relatively insensitive method (see above), but provides a fast and simple means of obtaining estimates of protein concentration over a rather large range of concentration (Itzhaki and Gill, 1964).

BIURET REAGENT

1. Combine 1.50 g $CuSO_4 \cdot 5\ H_2O$, 6.00 g sodium potassium tartrate, and 500 ml distilled water in a beaker and stir.
2. Add while stirring 300 ml of 10% NaOH (w/v).
3. Transfer to a 1-liter volumetric flask and bring to 1 liter with distilled water.
4. Transfer to a *plastic* bottle for storage. The reagent should be a deep blue color and is stable for about a year at room temperature.

Note: Biuret reagent can also be purchased from Sigma Chemicals for a reasonable cost. If the reagent, either prepared or purchased, shows signs of precipitation it should be discarded, and fresh reagent prepared.

BIURET ASSAY FOR SOLUBLE PROTEINS

Materials

1. Biruet reagent
2. Disposable cuvettes

3. Distilled water
4. 1 ml of 10 mg/ml bovine serum albumin (BSA)*
5. Closed end capillary tubes
6. An absorption spectrometer.

*BSA can be purchased from several suppliers, such as Sigma Chemical and CalBiochem as either a bulk solid, preweighted ampules of solid protein, or a concentrated solution of known protein mass per ml. My laboratory has found that it is convenient to prepare a large volume (say 100 ml) of a 10 mg/ml solution by carefully weighing out a gram of solid BSA and diluting this to 100 ml with distilled water (in a volumetric flask). The resulting solution is then divided into 1.0 ml aliquots which are stored in cryotubes. These are then flash-frozen by immersion into a dry ice/ethanol bath and stored at −20° C. The protein can be stored for more than a year under these conditions.

Procedure

1. To each of eight disposable cuvettes add the following reagents according to table 3.2:
2. To tubes 6–8 and 50 μl of your protein sample. Mix the contents of each tube well by using the closed end of the capillary tube as a stirring rod.
3. Next add 2.0 ml of the Biuret reagent to each tube, and mix well.
4. Let the tubes incubate at room temperature for 30–45 min, then read the absorbance of each tube at 540 nm.

Table 3.2 Experimental set up for the Biuret assay.

Tube Number	Water	10mg/ml BSA	Effective BSA Concentration (mg/ml)
1	500 μl	0	0
2	400 μl	100 μl	20
3	300 μl	200 μl	40
4	200 μl	300 μl	60
5	100 μl	400 μl	80
6	450 μl	0	unknown
7	450 μl	0	unknown
8	450 μl	0	unknown

5. For tubes 1–5 plot the absorbance at 540 nm as a function of effective BSA concentration, and calculate the best fit straight line from the data. Then, using the average absorbance for the three samples of your unknown, read the concentration of your sample from the plot.

A typical example of the standard plot resulting from this type of assay is shown in figure 3.1.

Modified Biuret for Turbid or Insoluble Samples

Because of its wide linear concentration range, the Biuret assay is often used for analysis of initial cell lysates, or membrane fragments at the early stages of a protein purification scheme. A potential problem in these cases is the effect of sample turbidity on the apparent absorbance reading obtained. To overcome this source of error, one can modify the Biruet assay by reducing the amount of water in each sample by 100 μl,

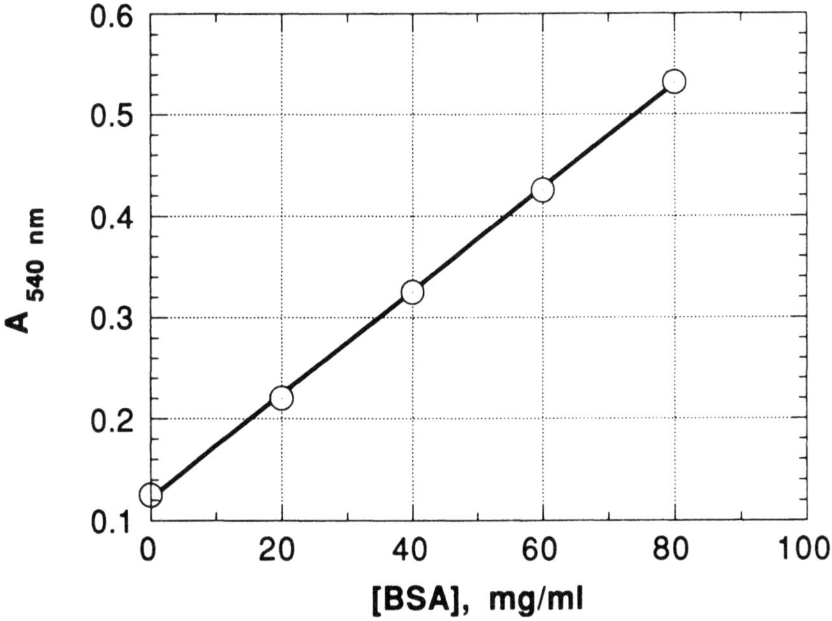

Figure 3.1 Typical results from the Biuret assay for protein concentration. Here the absorbance at 540 nm is plotted as a function of the concentration of bovine serum albumin (BSA) in the samples.

and replacing this volume with 100 μl of a 5% (w/v) solution of deoxycholate made up in 0.1 M KOH. Deoxycholate is a powerful detergent that helps to clarify solutions containing protein aggregates, cell membranes, and integral membrane proteins (that are not soluble in the absence of detergents). The deoxycholate itself can cause some color formation, but the interference is minimal. It is important, however, that the same amount of deoxycholate be present in all of the standards and unknowns used in the assay.

Lowry Assay (Useful Detection Range: 1 to 300 μg/ml)

This method is similar to the Biuret assay, but provides a much more intense color. Therefore, the method is more sensitive than the Biuret and can be used with lower concentrations of protein (Lowry et al., 1951).

LOWRY REAGENTS

Reagent A: Dissolve 100 g of Na_2CO_3 in 1 liter of 0.5 N NaOH.

Reagent B: Dissolve 1 g $CuSO_4 \cdot 5H_2O$ in 100 ml of distilled water.

Reagent C: Dissolve 2 g potassium tartrate in 100 ml of distilled water.

These reagents can be stored indefinitely.

Just before the assay combine 15 ml of reagent A, 0.75 ml reagent B, and 0.75 ml reagent C and mix well to form the Lowry reagent.

MATERIALS

1. Lowry reagent
2. Disposable cuvettes
3. Disposable 16 × 150 mm test tubes
4. 0.3 mg/ml BSA solution
5. 2N Folin-phenol reagent (from Sigma)
6. Vortex mixer.

PROCEDURE

1. In each of 10 tubes carefully pipette one of the following volumes of a 0.3 mg/ml solution of bovine serum albumin: 0, 0.1, 0.2, 0.3, 0.4, 0.5, 0.6, 0.7, 0.8, and 1.0 ml. In each of three other tubes, add 0.1 ml of your sample.

2. Bring the total volume of each tube to 1.0 ml with distilled water.
3. Add 1 ml of Lowry reagent to each of the tubes. Vortex the tubes to mix them thoroughly.
4. Incubate the tubes for 15 min at room temperature.
5. While the tubes are being incubated, add 5.0 ml 2N Folin-phenol reagent to 50 ml distilled water in a 125 ml Erlenmeyer flask. Mix the solution thoroughly.
6. At the conclusion of the incubation period, pipette 3.0 ml of the solution made in step 5 into each tube. Vortex the resulting solution immediately. It is very important that the addition and mixing be accomplished in as short a time as possible. The addition to and mixing of each tube should be completed before proceeding to the next.
7. Incubate the sample at room temperature for 45 min.
8. Determine the absorbance of each sample at 540 nm. In addition to 540 nm, the absorbance of the samples may be read at 750 nm for greater sensitivity.
9. Calculate the least squares best fit for the data, and the concentration of your samples as described above for the Biuret assay.

Note that the Lowry assay can be adapted for membrane protein in the same fashion described above for the Biuret assay.

A variety of reagents can interfere with both the Biuret and Lowry assays. Copper chelators, such as EDTA, must obviously be avoided. Detergents, lipids, and nucleic acids can also interfere, but can usually be compensated for; a very complete list of interfering compounds has been compiled by Peterson (1979).

Bradford Assay (Useful Detection Range: ≤ 25 μg/ml or 200 to 1400 μg/ml)

This assay is based on the binding of the protein-specific dye, Coomassie Brillant Blue to proteins (Bradford, 1976). Under the pH conditions of the assay, the dye is present in its cationic form, and does not absorb strongly at 595 nm. When the dye binds to a protein, there is a stabilization of the doubly protonated anionic form of the dye, which does absorb in the 595 nm region. As with the other assays described above, the details of protein-dye interactions here are not well understood. It has been empirically determined, however, that one requires peptides of nine residues or more that contain basic or aromatic residues to observe the color development upon interaction with the dye.

The reagents needed for the Bradford assay have been described in the literature and are commercially available from several sources. Bio-Rad Laboratories has performed a substantial study of the compatibility of this assay with various non-protein components of samples, and offers a protein assay kit, based on the Bradford assay, that is extremely convenient to use on a routine basis. The assay is described in detail in their bulletin number 1069 (available from Bio-Rad Laboratories, Richmond, California).

One of the major advantages of this assay is that, unlike the Biuret and Lowry assays, the Bradford assay is relatively insensitive to interferences from reagents that are commonly found in protein solutions. Bio-Rad Laboratories has compiled an extensive list of common reagents that do not interfere strongly with this assay; this list is provided in their bulletin number 1069.

The literature accompanying the Bio-Rad Bradford assay reagent describes two methods for performing the assay, a standard assay for samples in the range of 200–1,400 μg/ml, and a microassay for samples in the range of ≤ 25 μg/ml.

MATERIALS

1. Bio-Rad Dye Reagent Concentrate
2. Disposable cuvettes
3. Disposable 16 × 150 mm test tubes
4. Protein standard solutions
5. Vortex mixer.

PROCEDURE (STANDARD ASSAY)

1. Prepare a series of protein standard solutions of known concentration between 0.2 and 1.4 mg/ml (I recommend at least five samples within this range).
2. Dilute one volume of the Bio-Rad Dye Reagent Concentrate with four volumes of distilled water. Mix well and filter to remove any particulate matter.
3. Add 100 μl of each of the standard protein solutions, and your unknown to separate test tubes. To one test tube add 100 μl of buffer to serve as a zero protein blank.
4. To each test tube add 5.0 ml of the diluted assay reagent prepared in step 2. Vortex each tube to mix the sample well.

5. Incubate the samples for 5 min to 1 hour (be sure to incubate all tubes for an equivalent amount of time). Then transfer samples to the disposable cuvettes and read the absorbance of the sample at 595 nm.
6. Prepare a standard curve as previously described, and read off the concentration of your unknown.

PROCEDURE (MICROASSAY)

1. Prepare a series of protein standard solutions of known concentration between 1 and 25 μg/ml (I recommend at least five samples within this range).
2. Add 800 μl of each sample (standard, buffer blank, and unknowns) to a separate test tube.
3. To each tube add 200 μl of the Bio-Rad Dye Reagent Concentrate (do not dilute as in the standard assay), and vortex each tube to mix the samples well.
4. Incubate the samples as in the standard assay. Then transfer samples to the disposable cuvettes and read the absorbance of the sample at 595 nm.
5. Prepare a standard curve as previously described, and read off the concentration of your unknown.

SOME GENERAL COMMENTS ON COLORIMETRIC ASSAYS

One caution that must be noted concerning all of these colorimetric assays is that the choice of protein standards can have a dramatic effect on the concentration estimates obtained for unknowns. As an example of this, figure 3.2 illustrates the standard curves obtained for BSA and for bovine gamma globulin (IgG) using the Bradford method. While both proteins give linear responses to dye binding, the slopes of the lines are quite different. BSA tends to be more sensitive than proteins in general in these assays. In the procedures described above, we have used BSA for generating the standard plots. This is commonly done because of the low cost of BSA and its ready availability. However, it is clearly not always the best choice for very accurate concentration estimates. The best standard for any of these assays would be a stock solution, of known concentration, of the target protein itself. When this is not feasible, one must empirically determine whether one is better off using BSA or IgG as a primary standard for these methods.

Another point that should be made is that I have recommended the

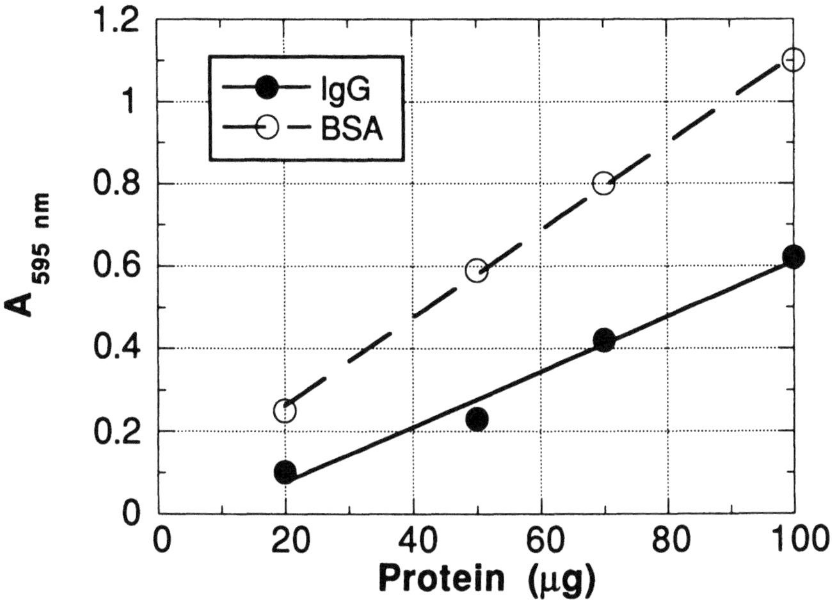

Figure 3.2 Comparison of the response observed for bovine serum albumin (BSA) and bovine immunoglobulin G (IgG) in the Bradford assay for protein concentration. Here the absorbance at 595 nm is plotted as a function of protein concentration for samples after incubation with Coomassie Brilliant Blue (see text for details).

use of disposable plastic cuvettes in all of the colorimetric assays. The reason for this is that some of the reagents (particularly Coomassie Brilliant Blue) adhere to glass and quartz surfaces. Thus, if glass or quartz cuvettes are to be used with these assays, they must be rigorously cleaned between samples. This procedure is inconvenient and time consuming. Given the low cost of disposable cuvettes, I think that they should be used universally for these assays. However, for those who need to use glass or quartz cuvettes, the method described at the end of this chapter should be used to clean them between samples.

SPECTROSCOPIC METHODS

The aromatic amino acids tyrosine and tryptophan are commonly found in proteins, and both give rise to π–π^* absorption bands in the vicinity of 280 nm (see table 3.3). For this reason, most proteins show a

Table 3.3 Wavelength maxima (λ max) and extinction coefficients (ε) for protein chromophores. From Fodor et al. (1989).

Chromophore	λ max (nm)	ε ($mM^{-1}cm^{-1}$)
Tyrosine	275	1.4
Tyrosinate	293	2.3
Tryptophan	280	5.6
Phenylalanine	257	0.2

strong absorption band in this region, which can be used to provide rough estimates of the concentration of protein in solution.

A general procedure for obtaining the absorption spectrum of a protein in solution is as follows:

1. Add to a clean, dry, quartz cuvette enough buffer to fill the cuvette at least three-quarters full.
2. Record the absorption spectrum between 250 and 350 nm of the buffer, and store this spectrum as the reference or baseline spectrum.
3. Remove the buffer and clean and dry the cuvette.
4. Filter your protein sample through a 0.22 μ filter, or centrifuge the sample to remove any particulate matter—including aggregated protein. Note that the removal of aggregated protein will lower the amount of total protein in the sample, but in most cases the aggregated protein should be removed from samples anyway. If it is desirable to obtain an estimate of total protein, aggregates can usually be dispersed by addition of 1% sodium dodecyl sulfate (SDS) or 6 M guanidine hydrochloride to the sample, and sonication for 5 minutes in a bath sonicator. Note, however, that these treatments lead to protein denaturation, and should thus be considered destructive to the sample.
5. Fill the cuvette with your sample of protein, and record the spectrum as above.
6. If your spectrometer does not automatically correct for the baseline, subtract the buffer spectrum from the spectrum of your sample to obtain the true absorption spectrum of your protein.

If this procedure is followed, most proteins will display a spectrum similar to that shown in figure 3.3, with a peak maximum at ca. 280 nm. At 260 nm the absorption of a pure protein sample is quite low compared to the maximum intensity at ca. 280 nm. The ratio of optical densities at these two wavelengths A_{280}/A_{260} should be on the order of 1.75 for a pure protein sample. If this is the case, a crude estimate of protein concentration can be obtained from the spectrum, based on the statistical average content of tyrosine and tryptophan in proteins.

$$[\text{Protein}],\ \text{mg/ml} \sim A_{280}$$

However, it must be clearly stated that this is a gross oversimplification, and the true extinction coefficient at 280 nm will vary considerably from protein to protein. Kirschenbaum (1978) has compiled an extensive list of the known extinction coefficient values for various proteins. If you are fortunate enough to have your target protein included in this list,

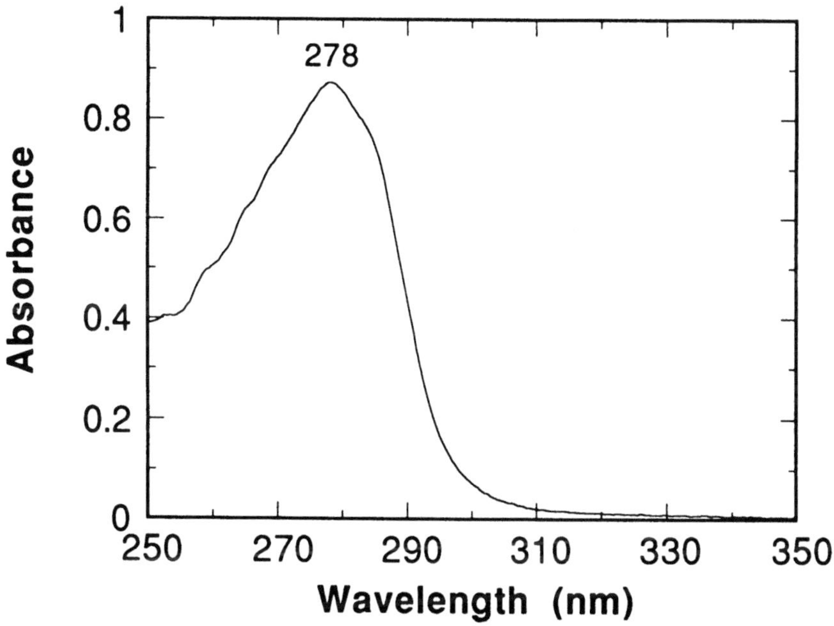

Figure 3.3 Typical UV absorption spectrum of a protein. Here the absorption spectrum of a 1 mg/ml solution of bovine serum albumin (BSA) in phosphate buffered saline is shown from 350 to 250 nm. The peak absorbance occurs at 278 nm, reflecting the combined contributions of tyrosine and tryptophan to this spectrum. The fine structure seen between 250 and 270 nm reflects some weak contributions from phenylalanine residues within the protein. See Chapter 9 for more details on the absorption properties of these amino acids.

then by all means use the correct literature value. It should also be noted, of course, that these literature values are for the pure forms of these proteins under specific solution conditions.

The presence of aggregated protein will have a tendency to cause light scattering that affects the overall shape of the absorption curve and tends to fill in the trough at 260 nm. This is why such aggregates need to be removed by filtration or centrifugation. If nucleic acids are present in the sample, they will absorb in the vicinity of 260 nm, and thus also have a tendency to fill in this trough region. A crude estimate of protein concentration can still be obtained in the presence of nucleic acids, if one measures the absorbance at 260 and 280 nm and uses the formula (Schleif and Wensink, 1981):

$$[\text{Protein}],\ \text{mg/ml} = 1.55A_{280} - 0.76A_{260}$$

An alternative formula for correcting protein spectra for nucleic acid contributions has been presented by Layne (1957):

$$[\text{Protein}],\ \text{mg/ml} = F \times A_{280}$$

where F is determined from the ratio A_{280}/A_{260} using table 3.4.

Of course, this is a very crude method since the relative amounts of tyrosine and tryptophan vary considerably from protein to protein. The

Table 3.4 Protein determination by ratio of optical densities A_{280}/A_{260} nm.

$A_{280/260}$	Nucleic acid %	F
1.75	0.0	1.116
1.52	0.5	1.054
1.36	1.0	0.994
1.16	2.0	0.899
1.10	3.0	0.814
0.939	4.0	0.743
0.874	5.0	0.682
0.822	6.0	0.632
0.784	7.0	0.585
0.753	8.0	0.545
0.730	9.0	0.508
0.705	10.0	0.478
0.645	14.0	0.377
0.585	20.0	0.278

estimates derived in this way should therefore only be used to obtain a rough idea of the amount of protein present.

If one knows the amino acid sequence of a protein, a better estimate of molar absorbance, ε_{280}, can be obtained by multiplying the values from table 3.3 for the individual amino acids by their copy number in the protein

At 280 nm:

$$\varepsilon_{\text{protein}}(\text{mM}^{-1}\ \text{cm}^{-1}) = (n \times 1.4) + (m \times 5.6)$$

where n = number of tyrosine residues
m = number of tryptophan residues

The problem with this method is that it does not take into account the environmental influences on the extinction coefficients of the amino acids. Estimates by this method are probably good to ± 10%.

Tyrosinate Difference Spectral Method

A much more accurate method for determining protein concentration is based on the changes in absorptivity of tyrosine residues when they are ionized at high pH (Fodor et al., 1989; Demchenko, 1986). The procedure takes advantage of the facts that (1) the difference extinction coefficient for tyrosinate minus tyrosine has been empirically determined to be relatively constant for a large number of proteins, and (2) the $\pi-\pi^*$ transitions of tryptophan, the other major contributor to protein near-UV absorption, are not affected by raising the solution pH to ca. 12.

MATERIALS

1. 0.1M phosphate buffered saline, pH 7.4
2. 0.1 M phosphate buffered saline, pH 12.0
3. Two UV-quality quartz semimicro cuvettes
4. A spectrophotometer capable of accurate measurements between 250 and 350 nm
5. A concentrated stock solution of protein.

PROCEDURE

1. Turn on the spectrophotometer and allow at least 30 min for the deuterium source to warm up.
2. Pipette 900 μl of pH 7.4 buffer into one cuvette and 900 μl of pH 12.0 buffer into another cuvette. Place the pH 12.0 cuvette in the sample

cell holder of the spectrophotometer, and place the pH 7.4 cuvette in the reference cell holder. With these cells in place, run a baseline reading from 350 to 250 nm and store this scan as the baseline.

3. Add 100 μl of the protein sample to each cuvette and mix carefully. Then run the absorption difference spectrum.
4. A large positive peak at ca. 295 nm and a small negative valley at ca. 275 nm should be observed. These result from the absorption maxima of tyrosinate and tyrosine in the sample and reference cells, respectively. A typical difference spectrum is shown in figure 3.4.

Calculations

The concentration of protein in your stock solution, in mM units, is determined from the following equation:

$$[\text{Protein}] = (\Delta A_{295\text{nm}}/2.330 \times N) \times 10$$

where: [Protein] = the protein concentration of your stock solution in mM units.

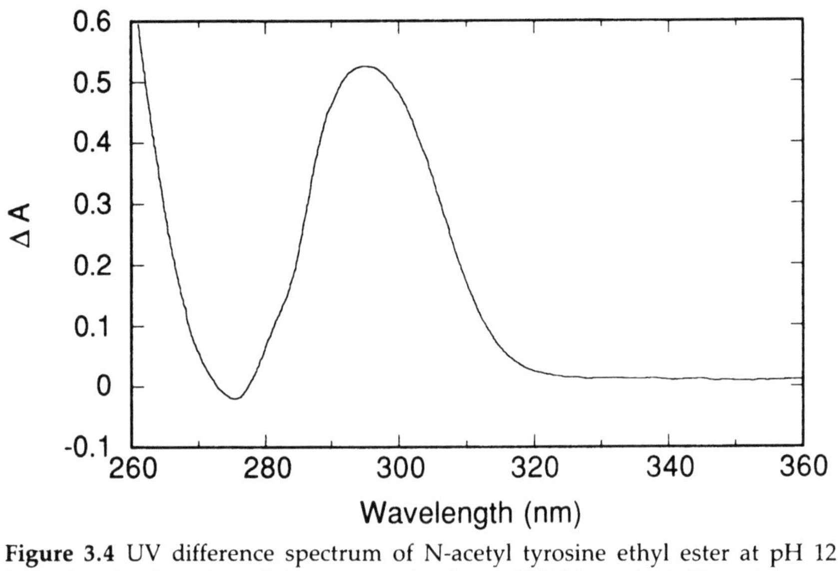

Figure 3.4 UV difference spectrum of N-acetyl tyrosine ethyl ester at pH 12 (tyrosinate) minus that of the same molecule at pH 7.0 (tyrosine). The minimum occurs at ca. 275 nm and the maximum at ca. 295 nm.

ΔA_{295nm} = the absorption at 295 nm from the pH 12.0 − pH 7.4 difference spectrum.
2.330 = the millimolar difference extinction coefficient for tyrosinate minus tyrosine.
N = the number of tyrosyl residue in the protein.
10 = the dilution factor (100 μl plus 900 μl = 1/10 dilution).

This procedure has been found to give very accurate concentration values for a large variety of proteins. The assay is not interfered with by other protein components, such as tryptophan, and is also free from interference by most common buffer components. For example, Smith et al. (1993) have used this method to quantitate the concentration of solutions of acidic fibroblast growth factor (aFGF) under various conditions. aFGF is a small protein that contains both tyrosine and tryptophan residues within its amino acid sequence. To determine if the presence of large quantities of tryptophan would interfere with this assay, Smith et al. (1993) examined the difference spectrum of a solution of N-acetyl tyrosine ethyl ester (Tyr) and an 8:1 molar mixture of N-acetyl tryptophan ethyl ester and N-acetyl tyrosine ethyl ester (8:1 Trp:Tyr). As seen in figure 3.5, the presence of excess tryptophan had no effect on the accuracy of the estimates of tyrosine concentration for these solutions. In the presence or absence of tryptophan one observes a linear response at $\Delta A_{295\ nm}$ to tyrosine concentration; the slope of this linear response is 2.330 for both solutions, the expected difference extinction coefficient for tyrosinate minus tyrosine. aFGF contains eight tyrosine residues within its amino acid sequence. One would thus expect that the difference extinction coefficient for this protein would be eight times that of tyrosine (e.g., 2.330 × 8 = 18.64 $mM^{-1}cm^{-1}$). Figure 3.6 illustrates the relationship between $\Delta A_{295\ nm}$ and the protein concentration for aFGF. As expected, the $\Delta A_{295\ nm}$ tracks linearly with protein concentration. The slope of this line is 18.65 $mM^{-1}cm^{-1}$, in excellent agreement with the expected values based on tyrosine itself. aFGF is one member of a family of growth factors whose native conformation is greatly stabilized by binding to the polyanion heparin. Quantitation of protein concentration for aFGF solutions containing heparin proved to be problematic using most conventional assays. In contrast, however, the tyrosinate difference spectrum method allowed accurate estimates of aFGF concentration even in the presence of the polyanion additive, as illustrated in figure 3.6. Smith et al. (1993) went on to test the accuracy of this assay for a number of growth factors of varying molecular weight and tyrosine content. These estimates were compared to those obtained by total amino acid analysis. In all cases the difference spectrum method proved extremely accurate

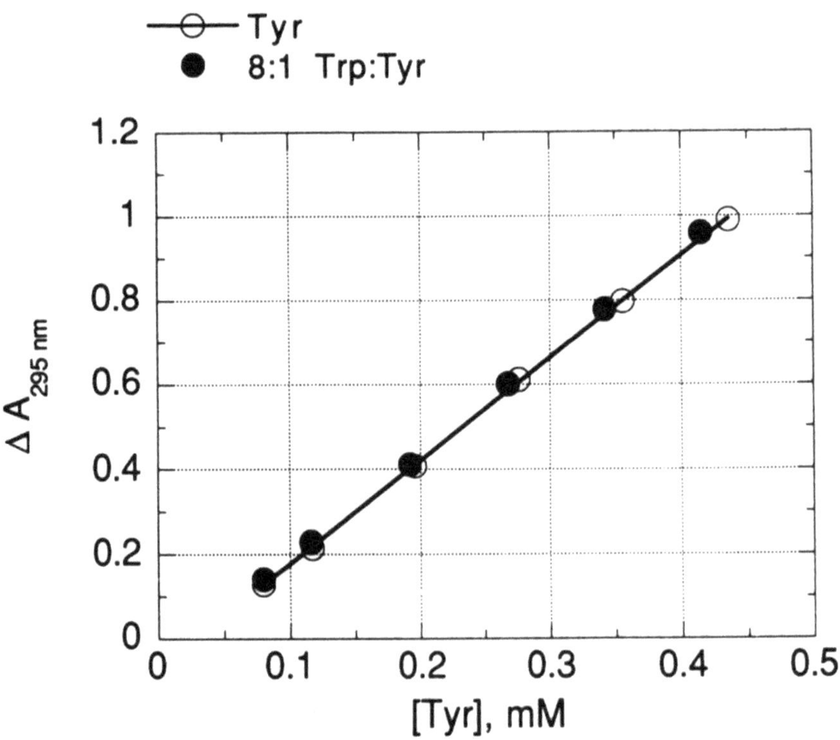

Figure 3.5 UV difference absorption at 295 nm ($\Delta A_{295\ nm}$) for samples at pH 12 minus pH 7 as a function of tyrosine concentration for solutions of N-acetyl tyrosine ethyl ester (open circles), and an 8:1 molar mixture of N-acetyl tryptophan ethyl ester and N-acetyl tyrosine ethyl ester (solid circles). The line drawn through the data represents the least squares best fit of the data to a linear equation.

and convenient. The analysis is accurate up to a ΔA_{295nm} value of ca. 1.0, beyond which one observes deviations from Beer's law. The only major drawback to this method is that it is destructive; a small amount of one's stock of protein will be irreversibly denatured by the high pH conditions of the assay.

SOME GENERAL COMMENTS ON ABSORPTION SPECTROSCOPY

When working in the ultraviolet region of the spectrum (below 350 nm), it is absolutely necessary that one work with high quality quartz cuvettes. Glass and plastic cuvettes absorb light strongly below about 350 nm, and are thus useless in this spectral region. Quartz cuvettes can be purchased from a number of suppliers, including many of the large scientific supply companies. Companies that specialize in manufacturing optical cells (such as Hellma, Precision Cells, and Spectrocell) offer a wide variety of high quality cell types. In particular, reduced volume, self-masking cells are very useful when samples are available in limited supply. For example, Hellma now offers an ultraviolet quality quartz cuvette that requires only 45 μl of sample to fill. My laboratory has used these cells for ultraviolet and visible absorption studies, as well as for protein fluorescence studies, with excellent results.

Optical cells also come in various pathlengths. As the reader may already know, the absorbance of any sample is related to its concentration, extinction coefficient, *and* the pathlength that the optical beam

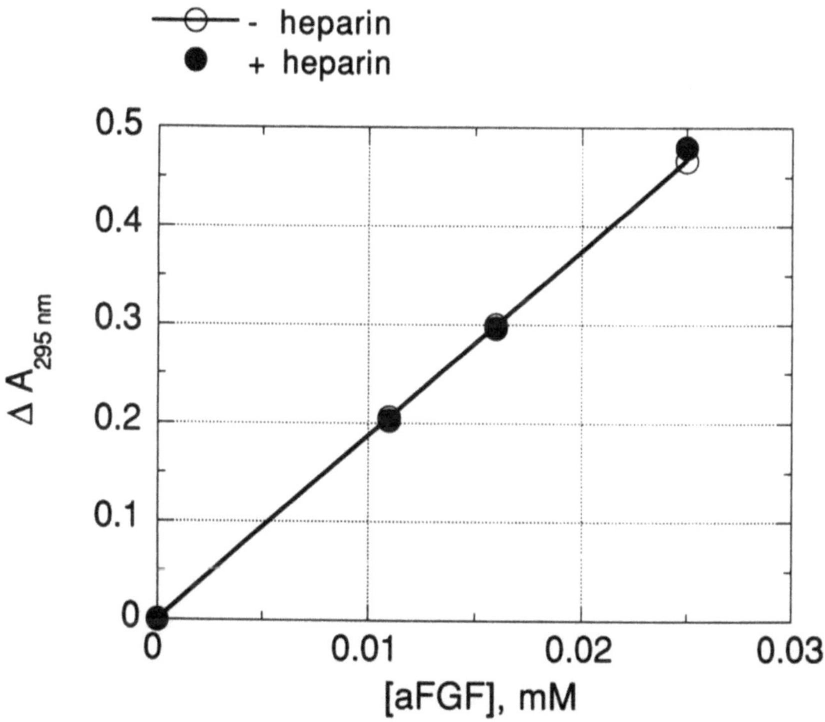

Figure 3.6 UV difference absorption at 295 nm ($\Delta A_{295\ nm}$) for samples at pH 12 minus pH 7 as a function of acidic fibroblast growth factor (aFGF) concentration for solutions of aFGF in the presence (solid circles) and absence (open circles) of heparin.

traverses through the sample. These are related by the well-known Beer's law:

$$A = \varepsilon c l$$

where A is the sample absorbance at a particular wavelength, ε is the molecular extinction coefficient for the sample at that wavelength, and l is the pathlength. In all of the calculations described above, and throughout the rest of this book, we shall assume that the pathlength is 1.0 cm. If a different pathlength cell is employed, one must correct the calculations accordingly.

No matter what type of optical cell one uses, it is necessary to ensure that the cell is rigorously clean before use. Optical cells should always be handled with latex gloves, and one should never touch the optical windows. Never use a metal or glass device for removing samples from a cuvette, as these may scratch the optical surfaces. As noted in Chapter 2, proteins adhere well to glass and quartz surfaces, as do some of the reagents described in this chapter (e.g., Coomassie Blue). To clean cells properly, it is useful to employ a commercially available cuvette washer, as illustrated in figure 3.7. This apparatus attaches to the top of a filtration flask, and allows one to force a stream of solution into the cell when vacuum is applied to the flask. These units are available from most of the major scientific supply houses, and from most cuvette manufacturers. The following procedure is recommended for cleaning cuvettes to be used with proteins:

1. Remove sample from the cuvette and fill the cuvette with a 1% solution of sodium dodecyl sulfate. Incubate at room temperature for 30 min or more.
2. Empty cuvette and place it into the cuvette washer. Apply vacuum to the flask.
3. Rinse the cuvette with a steady stream of 0.1 N HCl for about 10 seconds.
4. Rinse the cuvette with copious amounts of distilled water.
5. Rinse the cuvette briefly with HPLC grade methanol, and allow to dry *completely* by continued application of vacuum. Note that it is important to evaporate all of the methanol from the cuvette, since this solvent can be denaturing to proteins.
6. Carefully wrap the cleaned and dried cuvettes in lens paper (Kodak) and store them at room temperature in a location where they will be free from dust and moisture.

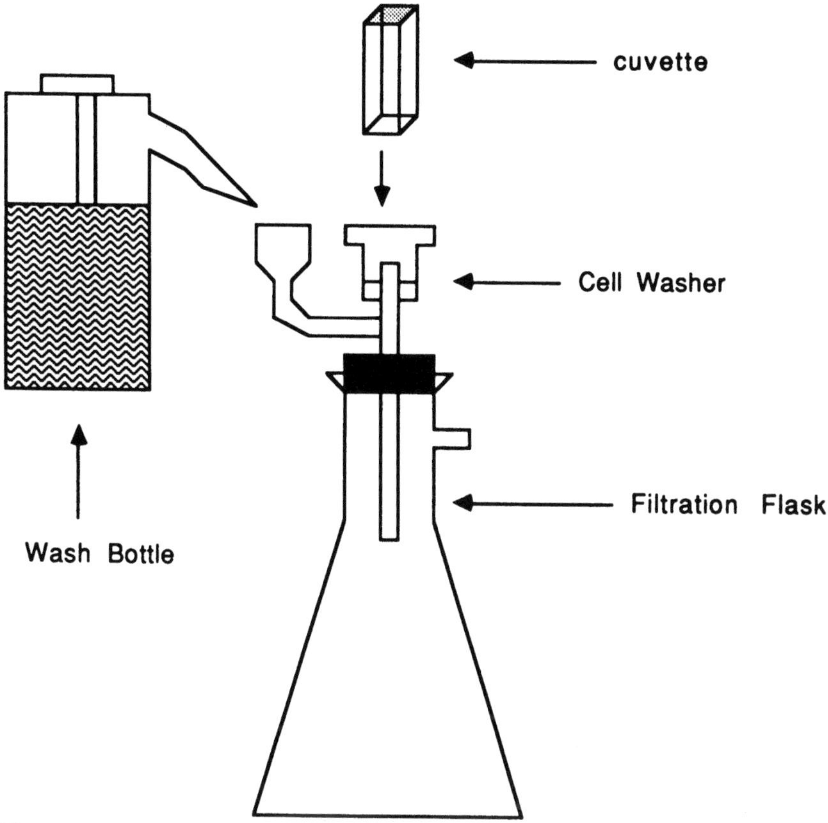

Figure 3.7 Schematic diagram of an optical cell washer used for cleaning cuvettes. Such devices are available from most of the major scientific supply houses.

References

Bradford, M. M. (1976) *Analyt. Biochem.*, **72,** 248.

Darbre, A. (1986) *Practical Protein Biochemistry: A Handbook*, Wiley, New York.

Demchenko, A. P. (1986) *Ultraviolet Spectroscopy of Proteins*, Springer-Verlag, New York.

Fodor, S. P. A.; Copeland, R. A.; Grygon, C. A.; and Spiro, T. G. (1989) *J. Am. Chem. Soc. USA*, **111,** 5509.

Itzhaki, R. F., and Gill, D. M. (1964) *Analyt. Biochem.* **9,** 401.

Kirschenbaum, D. M. (1978) *Analyt. Biochem.*, **90,** 309.

Kjeldahl, J. Z. Z. (1883) *Analyt. Chem.*, **22,** 366.

Layne, E. (1957) *Meth. Enzymol.*, **3,** 447.

Lowry, O. H.; Rosenbrough, N. J.; Farr, A. L.; and Randall, R. J. (1951) *J. Biol. Chem.*, **193,** 265.

Peterson, G. L. (1979) *Analyt. Biochem.* **100,** 201.

Schleif, R. F., and Wensink, P. C. (1981) *Practical Methods in Molecular Biology*, Springer-Verlag, New York.

Smith, T. S., Will, M. L., Cohen, H. P., and Copeland, R. A. (1993) *Analyt. Biochem.*, submitted for publication.

4

Electrophoretic and Chromatographic Methods for Assessing Protein Purity

ELECTROPHORESIS

Electrophoretic methods are widely used in protein science to determine sample purity, molecular weight, and sometime isoelectric point. In this chapter we shall review the most commonly used electrophoretic methods that are applied to proteins.

Native and denaturing gel electrophoresis have historically been used for assessing sample purity and molecular weight under conditions that do not or do lead to protein subunit dissociation, respectively. Isoelectric focusing is used to determine the isoelectric point of proteins, and provides a second criterion for assessing protein purity. While these methods have been used for thirty years or more, they remain powerful and popular methods today. Capillary zone electrophoresis is a relatively new method tl:at combines the power of traditional electrophoretic methods with new HPLC-like technologies.

SDS-PAGE

Perhaps the most widely used technique in protein science is sodium dodecyl sulfate-polyacrylamide gel electrophoresis (SDS-PAGE). The basis of this method is that a charged molecule will migrate in an electric field at a rate that is determined by its size and charge. Here the electric field is applied across a sheet of a polymer (polyacrylamide) that acts as a barrier to molecular motion. Before entering the electric field, the protein is denatured (unfolded) by exposure to harsh conditions (e.g., heat, denaturing detergents, disulfide reducing agents, and sometimes chaotropic agents such as urea), and is coated with the anionic detergent, SDS. In the denatured state, most proteins bind SDS in a constant weight ratio, so that they end up possessing similar charge densities. Under these conditions, the migration rate of the proteins in the electric field is no longer dependent on the molecules' inherent charge, but rather is determined solely on the basis of molecular size (i.e., larger proteins will be more severely retarded from migration in the polymeric gel than will smaller proteins). Figure 4.1 shows a schematic diagram of a typical SDS-PAGE apparatus. Here the protein samples are loaded into wells at the top of the gel, which is in contact, through a buffer reservoir, with an attached cathode. The bottom of the gel is likewise connected to an attached anode. When current is applied, the SDS-coated proteins migrate to the bottom of the gel, under the influence of the applied electric field.

Polyacrylamide is the most commonly used polymer for gel electrophoresis of proteins. The gels themselves are formed by the polymerization of acrylamide by a free radical mechanism (Hames and Rickwood, 1990). The molecular weight resolution that is achieved by SDS-PAGE depends in part on the pore size of the polymeric gel. This in turn is a function of the percentage of acrylamide used in the preparation of the gel. The percentage of acrylamide used will depend on the molecular weight range over which one wants to fractionate the samples. Table 4.1 pro-

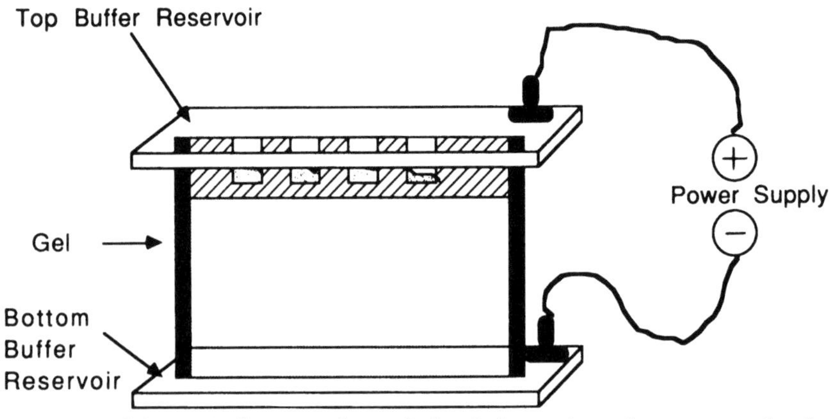

Figure 4.1 Schematic diagram of a typical gel electrophoresis apparatus for the separation of proteins in polyacrylamide.

Table 4.1 *Percentage of acrylamide to be used for resolving proteins of different molecular weights by* SDS-PAGE.

For Proteins of Molecular Weight Between:	Use This Percentage of Acrylamide:
200,000–60,000	5.0%
120,000–30,000	7.5%
75,000–18,000	10.0%
60,000–15,000	12.5%
45,000–12,000	15.0%

vides a guide to selecting the percentage of acrylamide to be used for fractionating proteins of different molecular weight ranges. In using this table, one must remember that multisubunit proteins will be resolved into the individual polypeptides in a denaturing gel. Thus, it is the molecular weights of the subunits, and not the native molecular weight of the assembly, that must be considered in choosing the acrylamide percentage.

Electrophoretic gel/buffer systems can be either homogeneous (continuous) or multiphasic (discontinuous). Homogeneous systems contain the same buffer ions and pH in the sample preparation, electrode buffer, and gel. Samples are loaded directly onto the revolving gel where separation occurs. In multiphasic buffer systems a stacking gel of different pH and/or buffer composition is employed to concentrate and sharpen sample constituents before they enter the resolving gel. The electrode buffer may also be of different pH and/or buffer composition. This discontinuity of pH and buffer composition results in higher resolution.

Apparatus for SDS-PAGE are available from a variety of manufacturers (e.g., Bio-Rad, Hoeffer, Pharmacia) in a number sizes and formates. Mini-gels are popular because they can be run quickly, and provide adequate resolution for many analytical purposes. However, for maximum resolving power, the larger formate gel systems are superior.

It is a common practice to include a mixture of proteins, of very well-defined molecular weights, in one or more lanes of the gel to serve as molecular weight standards (or markers). The molecular weight of one's target protein can then be determined by comparison with the relative mobilities of these standards, as described below. Premixed molecular weight standards can be purchased from many suppliers (e.g., Sigma, Bio-Rad, Hoefer, Pharmacia, Amersham). One can also purchase molecular weight standard kits in which the proteins have been predyed so that they can be visualized during the gel run. Amersham supplies a set of predyed molecular weight markers referred to as "Rainbow Markers," in which each protein is covalently bonded to a different color dye. My laboratory has found these markers to be extremely useful, because they allow one to watch the resolution of the protein bands as the gel is running (and thus detect any problems with the run), and because the different colored dyes eliminate any ambiguity as to which band corresponds to which molecular weight in the standard lane.

SDS-PAGE for Soluble Proteins

The protocol described below is a general one for soluble proteins. It has been adapted from a protocol described in Ausubel et al. (1989), as modified by my laboratory. This protocol is designed for a 0.75 mm × 14 cm × 14 cm slab gel. For gels of different dimensions, the volumes used must be adjusted.

MATERIALS

1. Protein molecular weight standards
2. 30% acrylamide/0.8% bisacrylamide
3. 4X Tris-C1/SDS, pH 8.8
4. 10% (w/v) ammonium persulfate
5. 10% (w/v) SDS
6. TEMED (N,N,N′,N′-tetramethylethylenediamine)
7. Isobutyl alcohol, H_2O-saturated
8. 4X Tris-Cl/SDS, pH 6.8
9. 2X SDS/sample buffer
10. 1X SDS/electrophoresis buffer
11. Bromphenol Blue solution
12. 0.75-mm spacers
13. 25 ml Erlenmeyer sidearm flasks
14. Teflon combs
15. 25 and 100 μl syringes with flat-tipped needles
16. 0.45-μm filters.

30% ACRYLAMIDE/0.8% BISACRYLAMIDE

Mix 30.0 g acrylamide and 0.8 g N,N′-methylene-bisacrylamide in a total volume of 100 ml H_2O. Filter the solution through a 0.45-μm filter and store at 4° C in the dark. Discard after 30 days, since acrylamide gradually hydrolyzes to acrylic acid and ammonia.

> ***CAUTION:*** *Acrylamide monomer is neurotoxic. Gloves should be worn while handling the solution, and the solution should not be pipetted by mouth.*

4X TRIS-CL/SDS, pH 6.8 (0.5 M TRIS-Cl CONTAINING 0.4% SDS, ALSO CALLED 4X UPPER TRIS)

Dissolve 6.05 g Tris base and 4 ml of 10% SDS in 40 ml H_2O. Adjust to pH 6.8 with 1 N HCl. Add distilled water to 100 ml total volume. Filter the solution through a 0.45-μm filter and store at 4° C.

TRIS-CL/SDS, pH 8.8 (1.5 M TRIS-Cl CONTAINING 0.4% SDS, ALSO CALLED 4X LOWER TRIS)

Dissolve 18.17 g Tris base and 4 ml of 10% SDS in 40 ml H_2O. Adjust to pH 8.8 with 1 N HCl. Add distilled water to 100 ml total volume. Filter the solution through a 0.45-μm filter and store at 4° C.

5X SDS/ELECTROPHORESIS BUFFER

Dissolve 15.1 g Tris base, 72.0 g glycine, and 5.0 g SDS in about 800 ml of H_2O. After the solutes are dissolved, bring the volume to 1.0 liter. Filter the solution through a 0.45-μm filter and store at 4° C. To prepare 1X SDS/electrophoresis buffer, dilute one volume of the above solution with four volumes of distilled water.

2X SDS/SAMPLE BUFFER

Mix 30 ml of 10% SDS, 10 ml glycerol, 5.0 ml 2-mercaptoethanol, 12.5 ml of 4X Tris-Cl/SDS, pH 6.8, and 5–10 mg Bromphenol Blue (enough to give a dark blue color to the solution). Bring the volume to 100 ml with distilled water. Divide into 1.0 ml aliquots, and store at −20° C.

POURING THE SEPARATING GEL

1. Assemble the mini-gel apparatus (e.g., Bio-Rad, Mini-Protean II, or other commercial unit) according to the manufacturer's detailed instructions. Make sure that the glass plates and other components

are rigorously clean and dry before assembly. Wear laboratory gloves when handling these components.

2. Mix the separating gel solution by adding, as defined in table 4.2 30% acrylamide/0.8% bisacrylamide, 4X Tris-Cl/SDS, pH 8.8, and water to a 25-ml sidearm flask. Degas the solution for 15 min by placing a rubber stopper in the opening of the flask and connecting a piece of vacuum tubing to the sidearm and to a vacuum line. Add 50 μl of 10% ammonium persulfate and 10 μl TEMED to the degassed solution and gently swirl to mix.

   CAUTION: Acrylamide is a potent neurotoxin. Avoid breathing in powder, ingestion, or contact with skin.

3. Using a Pasteur pipette, transfer the separating gel solution to the center of the sandwich along an edge of one of the spacers until the height of the solution in the sandwich is ~ 1.5 to 2 cm from the top of the shorter (front) glass plate (see figure 4.2). *Use the solution immediately; otherwise, it will polymerize in the flask.*
4. Using another Pasteur pipette, slowly cover the top of the gel with a layer (~3 mm thick) of isobutyl alcohol by gently squirting the isobutyl alcohol against the edge of one of the spacers. *Be careful not to disturb the gel interface.*
5. Allow the resolving gel to polymerize fully (usually 30 to 60 min).

POURING THE STACKING GEL

6. Pour off completely the layer of isobutyl alcohol. Rinse the separating gel top three times with distilled water.
7. Prepare a 5% stacking gel solution by adding 0.65 ml of 30% acrylamide/0.8% bisacrylamide, 1.25 ml of 4X Tris-Cl/SDS, pH 6.8, and

Table 4.2 Solution mixtures for preparing polyacrylamide gels.

	Final % of Acrylamide					
Stock to add (ml)	5	7.5	10	12	15	20
30% acrylamide/0.8% bisacrylamide	2.50	3.75	5.00	6.00	7.50	10.00
4X Tris-Cl/SDS, pH 8.8	3.75	3.75	3.75	3.75	3.75	3.75
Water	8.75	7.50	6.25	5.25	3.75	1.25
10% ammonium persulfate	0.05	0.05	0.05	0.05	0.05	0.05
TEMED	0.01	0.01	0.01	0.01	0.01	0.01

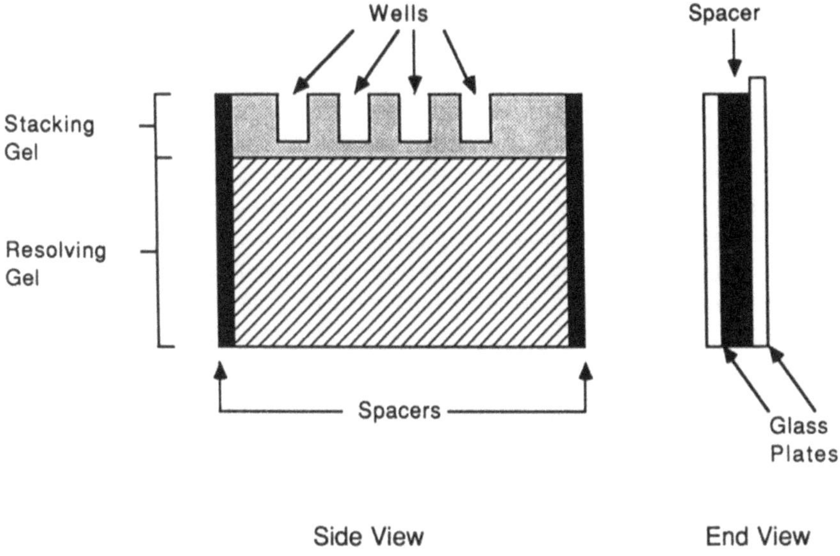

Figure 4.2 Side and end views of a discontinuous gel set up for protein electrophoresis. This type of set up is commonly used for SDS-PAGE as described in the text.

3.05 ml water to a 25-ml sidearm. flask. Degas the solution as above. Add 25 μl of 10% ammonium persulfate and 5 μl TEMED to the degassed solution and gently swirl to mix. *Use the solution immediately; otherwise, it will polymerize in the flask.*

8. Using a Pasteur pipette, slowly allow the stacking gel solution to trickle into the center of the sandwich along an edge of one of the spacers until the height of the solution in the sandwich is ~3 cm. *Be careful not to introduce air bubbles into the stacking gel.*
9. Insert a Teflon comb into the layer of stacking gel solution by placing one corner of the comb into the gel and slowly lowering the other corner in. Try to avoid trapping air bubbles in the gel while inserting the comb. Add additional stacking gel to completely fill the spaces in the comb, if necessary. *The Teflon combs are commercially available as blank, 3-well, 5-well, 10-well, 15-well, and 20-well.*
10. Allow the stacking gel solution to polymerize 30 to 45 min at room temperature.

LOADING THE GEL

11. Dilute the protein to be analyzed at least 1:1 (v/v) with 2X SDS/sample buffer in a microcentrifuge tube and boil for 1 min at 100° C. This is

easily done by filling a 250 ml beaker with water and covering the top with aluminum foil. A pencil is then used to poke small holes in the aluminum foil, and the beaker is placed on a hot plate to heat. When the water begins boiling, insert the microcentrifuge tube into the pencil holes until the entire sample is under the water. The centrifuge tubes should be capped to prevent evaporation of the sample solution. A small pin hole can be made in the tube cap with a syringe needle to relieve the pressure build up during boiling. If the sample is a precipitated protein pellet, dissolve the protein to be separated in 50 to 100 μl SDS/sample buffer and boil 1 min at 100° C. After boiling, cool the sample to room temperature, and centrifuge for a few seconds to remove particulate matter.

12. Carefully remove the Teflon comb without tearing the edges of the polyacrylamide wells. This is best done by grasping the comb on each end and smoothly pulling straight up out of the gel.

13. Using a Pasteur pipette, fill the wells with 1x SDS/electrophoresis buffer. *If well walls are not upright they can be manipulated with a flat-tipped needle attached to a syringe.*

14. Place the upper buffer chamber over the sandwich and lock the upper buffer chamber to the sandwich. Pour 1X SDS/electrophoresis buffer into the lower buffer chamber. Place the sandwich attached to the upper buffer chamber into the lower buffer chamber.

15. Partially fill the upper buffer chamber with 1X SDS/electrophoresis buffer so that the sample wells of the stacking gel are filled with buffer.

16. Using a 25- or 100-μl Hamilton syringe with a flat-tipped needle, load the protein sample(s) into the wells by carefully applying the sample as a thin layer at the bottom of the well(s), as shown in figure 4.3. The glycerol in the sample buffer gives the sample greater density than the buffer in the well, so the sample should sink to the bottom of the well if applied slowly. If one adds sample too rapidly or allows an air bubble to enter the well, the sample can escape the well and potentially contaminate other wells of the gel. If a Hamilton syringe is used to load samples, it must be washed with copious amounts of 1X SDS/sample buffer after every sample, to avoid the possibility of cross contamination. An alternative method for loading the sample is to use gel loading tips with a pipetter. These disposable tips greatly reduce the chances of cross contamination, since a fresh tip is used for each sample well. When possible, it is a good idea to leave a well

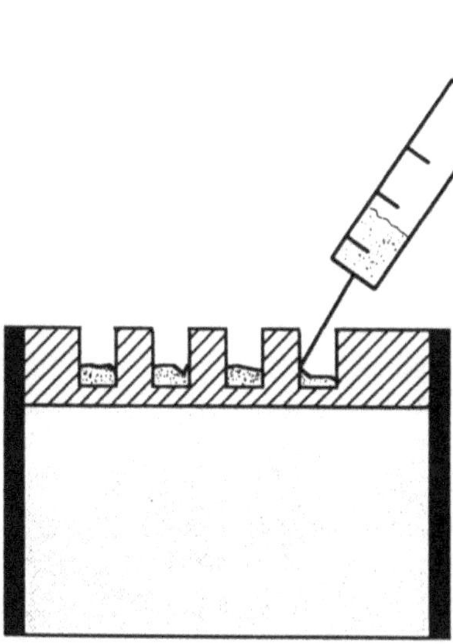

Figure 4.3 Loading of protein samples into the wells of a stacking gel. Here a Hamilton syringe is used to load the samples. Note that the tip of the needle is placed near the bottom corner of the well. Sample is released from the syringe **slowly** to form a continuous layer at the bottom of the well. Air bubbles must be rigorously avoided during loading to avoid escape of samples from the wells and hence contamination of adjacent wells. Gel loading tips, in conjunction with a pipetter, are a modern and convenient alternative to the use of a Hamilton syringe (see text).

empty between each sample well. When this is done, the empty wells should be loaded with 1X SDS/sample buffer (this will help prevent bleeding of sample across lanes of the gel and provide a uniform dye front for tracking the progress of the electrophoresis).

17. Fill the remainder of the upper buffer chamber with additional 1X SDS/electrophoresis buffer.

RUNNING THE GEL

18. Connect the power supply to the anode and cathode of the gel apparatus and run at 125 v.
19. After the Bromphenol Blue tracking dye has reached the bottom of the separating gel, disconnect the power supply. The total run time for a gel is ~4 h.

DISASSEMBLING THE GEL

20. Remove the upper buffer chamber and the attached sandwich.
21. Orient the gel so that the order of the sample well(s) is known, remove the sandwich from the upper buffer chamber, and lay the sandwich on a sheet of absorbent paper or paper towels.
22. Carefully slide the spacers out from the edge of the sandwich along its entire length. Insert a spatula between the glass plates at one corner where the spacer was, and gently pry the two plates apart. The gel will adhere to one of the plates. With the gel oriented so that the order of sample wells is known, cut off a small section of one of the lower corners with a spatula to mark the gel orientation.
23. Carefully remove the gel from the lower plate. This can be done by placing the plate with the gel attached into a shallow dish of fixing agent or dye (see below) and swishing the plate (gel side down, immersed in the solution) gently until the gel partially separates. Use a spatula to gently remove the rest of the gel from the plate. The gel can be stained with Coomassie Brilliant Blue or silver stain (see below), or transferred to a membrane filter for Western blotting (See Chapter 5).

Visualizing Protein Bands by Staining

After running the gel one needs some way of visualizing the protein bands. The most common way to do this is by staining the gel with a protein-binding dye. Two useful methods are Coomassie Blue Staining and Silver Staining.

In my laboratory we use a variation of the standard Coomassie Blue method, which is based on using Coomassie Brilliant Blue G-Perchloric Acid solution. This reagent can be purchased from Sigma Chemical Co. (St. Louis, MO; catalog no. B 8772). The advantages of this method is that it is simple to use and gives virtually no background staining to the gel, making visualizing of protein bands easy.

COOMASSIE BLUE VARIATION PROCEDURE

1. Place the gel in a small plastic box and cover with Brilliant Blue G-Perchloric Acid solution. Agitate slowly for 1 h or more (if convenient, the gel can remain in the staining solution overnight) on a rotary rocker.
2. Pour off the staining solution and cover the gel with a solution of methanol/water (45%/55%,v/v). Agitate slowly for ~30 min.

3. Discard destaining solution and replace with fresh solution. Repeat until the gel is clear (no color) except for the protein bands.

This procedure works well when one has relatively large quantities of protein. The detection limit is 0.3–1.0 μg protein/band. Figure 4.4 illustrates the results of a typical SDS-PAGE run after staining with Coomassie Brilliant Blue G-Perchloric Acid.

Silver staining is an alternative which provides a much lower detection limit, ~2–5 ng protein/band.

MATERIALS

DESTAINING SOLUTION

5% (v/v) methanol

7% (v/v) acetic acid

88% (v/v) H_2O

DEVELOPING SOLUTION

0.5 g sodium citrate

0.5 ml 37% formaldehyde solution (Kodak)

H_2O to 100 ml

FIXING SOLUTION

50% (v/v) methanol

10% (v/v) acetic acid

40% (v/v) H_2O

SILVER NITRATE SOLUTION

Add 3.5 ml concentrated NH_4OH (~30%) to 42 ml of 0.36% NaOH and bring the volume to 200 ml with H_2O. Mix with a magnetic stirrer and slowly add 8 ml of 19.4% (1.6 g/8 ml) silver nitrate.

If the solution is cloudy, carefully add NH_4OH until it clears. The solution should be used within 20 min.

SILVER STAINING PROCEDURE

Agitate gel slowly during steps 1 through 4.

1. Place the polyacrylamide gel in a plastic box on an orbital shaker and add fixing solution. Incubate for 30 min.

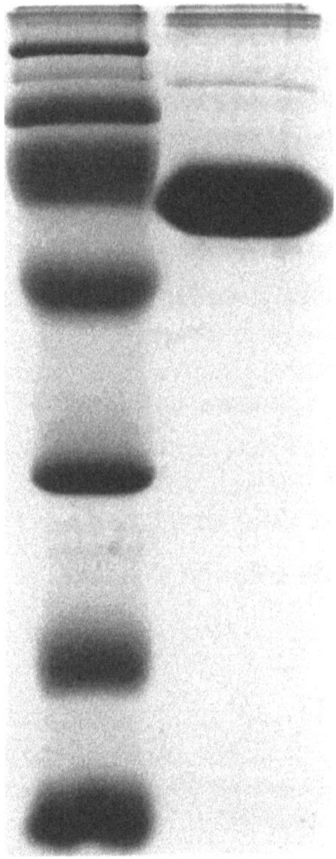

Figure 4.4 SDS-PAGE of proteins. A Bio-Rad Mini-Protean II apparatus was used to cast and run the gel. The left lane contains molecular weight standards (Amersham Rainbow Markers) and the right lane contains a putatively pure sample of bovine serum albumin (BSA). The samples were prepared in 1X sample buffer and boiled for 1 min as described in the text. They were loaded into wells in a 5% acrylamide stacking gel and resolved with a 10% resolving gel. The proteins were stained with Coomassie Brilliant Blue G-Perchloric Acid and destained as described in the text. Data provided by Carol Wilson, Department of Biochemistry, The University of Chicago.

2. Decant the fixing solution, and fix the gel in destaining solution for at least 60 min.
3. Decant the destain, and cover the gel with 10% glutaraldehyde. Agitate for 30 min.
4. Wash the gel four times with water, for at least 30 min each wash.
5. Stain the gel with silver nitrate solution for 15 min with vigorous shaking

   CAUTION: Dispose of the silver solution immediately, since it becomes explosive upon drying.

6. Transfer the gel to another plastic box and wash five times with water (exactly 1 min for each wash).
7. Prepare the developer by diluting 25 ml developing solution with 500 ml water. Transfer the gel to another plastic box, add developer, and shake vigorously until the bands appear as intense as desired. If the developer turns brown, change to fresh developer.
8. Transfer to Kodak Rapid Fix for 5 min.
9. Wash gel exhaustively in water to remove Rapid Fix.

Some Comments on Coomassie and Silver Staining

As the reader can probably guess from the preceding procedures, Coomassie staining is considerably less complicated than silver staining. An additional advantage of Coomassie staining is that different proteins tend to stain to the same extent with this dye, thus making it possible to roughly quantitate the relative amounts of different proteins by comparison of the intensities of bands via densitometry (Hames and Rickwood, 1990). Silver staining, on the other hand, provides far greater sensitivity than Coomassie Blue, but shows considerably more protein to protein variation in the degree of staining. For this reason, silver staining is useful for detecting the presence of minor contaminating proteins in a sample, but is not recommended for quantitation. Whether one uses Cooomassie or silver staining, if the goal is to search for minor contaminating proteins, it is wise to run several lanes on the same gel with increasing amounts of the target protein. At the highest loading concentrations, the target protein band may be too dark and broad for quantitation (referred to as overloading), but minor bands will be accentuated in these lanes. Several commercial kits are now available for silver staining that eliminate the need for the researcher to prepare all of the reagents for this method.

Other Visualization Methods

The procedures outlined above have been detailed because they have been found to provide consistent and convenient staining of a wide variety of proteins. Other methods for protein visualization have been documented elsewhere. These include copper staining, colloidal gold staining, and autoradiography of radiolabeled proteins (Hames and Rickwood, 1990).

Molecular Weight Determination

The molecular weight of a target protein can be roughly determined from its relative mobility on the gel, if molecular weight standards are included in a separate lane of the same gel. The relative mobility of a protein band in a gel R_f is defined as the distance of migration of that band (from the top of the resolving gel, measured in cm with a ruler) divided by the distance of migration of the dye front. A plot of the log (molecular weight) as a function of R_f for the molecular weight standards of a typical gel is shown in figure 4.5. From this linear relationship, and the measured R_f value of the target protein, one can estimate the molecular weight of the target protein. It should be noted, however, that these estimates are not exact. The mobility of a protein in the gel is largely, but not entirely, determined by its molecular weight. For example, glycosylation, or other post-translational modifications of proteins can alter their apparent polypeptide molecular weights as determined from gel electrophoresis. For most single subunit proteins, however, the molecular weights determined by this method are within $\pm$ 10% of the true molecular weights.

Record Keeping

Once the gel has been run and stained one needs some way of permanently recording the results. The most common way to do this is to photograph the gel and store the photo in a laboratory note book. The gel itself can be stored long term by placing it in a "ziplock baggie" with a small volume of methanol/water.

Another way to store the gel is to dry it in a gel drying apparatus. These are available in a variety of styles and prices. A simple, easy, and inexpensive way of drying gels is to sandwich them between two sheets

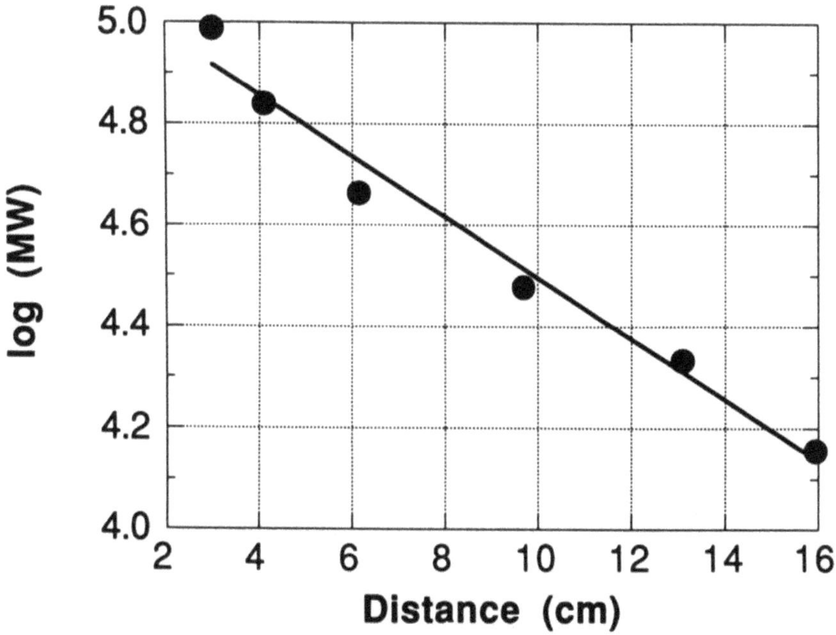

Figure 4.5 Plot of log (molecular weight) of proteins as a function of their migration distance in SDS-PAGE. The data presented here are for the molecular weight marker proteins in the left lane of the gel shown in figure 4.4. This type of graph is often referred to as a Ferguson plot.

of hydrated cellulose, and then place the sandwich in a rigid frame until it is dry. Commercial devices using this method are available from several sources. Ann Arbor Plastics (Ann Arbor, MI), for example, provides an inexpensive and effective system for gel drying.

SDS-Urea Gels for Membrane Proteins

Proteins of high hydrophobicity, such as integral membrane proteins, sometimes do not run well in conventional SDS gels. Two problems often encountered with these systems are incomplete subunit dissociation and poor solubility in the gel. To overcome these problems, high concentrations of urea can be included in the gel system. An additional problem that can be encountered with membrane proteins is interferences due to the detergents or phospholipids present in solution for protein solubilization. This problem can be easily overcome by TCA precipitation of the protein prior to resolubilization in SDS/sample buffer (see Chapter 2). The procedure for running SDS-urea gels is the same as described above, except that urea is contained in the stacking and resolving gel solutions. The following procedure, modified from Kadenbach et al. (1983), has worked well for membrane proteins in my laboratory.

2X MEMBRANE PROTEIN SAMPLE BUFFER

Mix 12.5 ml 4X Tris-Cl/SDS, pH 6.8, 50 ml of 10% SDS, 10 ml glycerol, 5.0 ml 2-mercaptoethanol, and 1 mg Bromphenol Blue. Adjust the pH to 6.8 with HCl, and bring the volume to 100 ml with distilled water. Divide into 1.0 ml aliquots, and store at −20° C.

POURING THE SEPARATING GEL

1. Combine 8 ml of 30% acrylamide/0.8% bisacrylamide, 7.5 ml of 1M Tris-Cl, pH 8.8, 0.2 ml of 10% (w/v) SDS, 2.6 ml glycerol, 4.3 g urea, and 2.7 ml distilled water in a beaker. Filter the solution through a 2.2 μ syringe filter.
2. Place the solution in a 50 ml vacuum flask, and degas as above.
3. Add 50 μl of 10% ammonium persulfate and 10 μl of TEMED. Swirl to mix, and load into gel apparatus as described above.

Note that this gel differs from those previously described in that it contains a higher percentage of SDS, and includes urea.

POURING THE STACKING GEL

1. Combine 2.5 ml of 30% acrylamide/0.8 bisacrylamide, 1.0 ml of 1M Tris-Cl, pH 6.8, 100 μl of 10% SDS, and 6.4 ml of distilled water in a beaker. Filter the solution through a 0.22μ syringe filter.
2. Place the solution in a vacuum flask and degas.
3. Add 30 μl of 10% ammonium persulfate and 14 μl of TEMED. Swirl to mix, and load into gel apparatus as above.

SAMPLE PREPARATION

1. Place a volume of sample corresponding to 5–10 μg (less for silver staining) of protein in a microcentrifuge tube. Add to this 10 volumes of cold 5% TCA to precipitate the protein. Let the sample incubate on ice for 15–30 min (this step helps to remove detergent and salts from the sample that might interfere with the electrophoresis).
2. Centrifuge samples in a microcentrifuge for 5–10 min.

3. Aspirate off the supernatant and wash the pellet with ethanol to remove excess TCA (see Chapter 2).
4. Resuspend the pellet in 10 μl of 4X Tris-Cl/SDS, pH 6.8 buffer, and 10 μl of 2X Membrane Protein Sample Buffer. Mix well and sonicate in a bath sonicator at 4°C for 30 min. Incubate for 30 to 60 min at 37° C.
5. Centrifuge for 10 seconds to remove any particulate matter, and load onto gel as above.

Nondenaturing Gels

There are times when one wishes to assess the purity and molecular weight of a target protein under conditions that do not lead to protein denaturation. For example, for multisubunit proteins, one may wish to determine the molecular weight of the native assembly rather than that of the component subunits. Likewise, one may ask if all of the bands seen in a denaturing gel correspond to true subunits of the protein, or if some of these are contaminating proteins. Nondenaturing gel electrophoresis is one method for addressing these issues.

Nondenaturing gels are prepared and run as described above for denaturing gels with the following exceptions:

1. The Tris buffers used for preparing the stacking and separating gels, and the electrophoresis running buffer are prepared as described previously *except that the SDS is left out for nondenaturing gels.*
2. The electrophoretic sample buffer is prepared as described previously *except that the SDS and the 2-mercaptoethanol are left out for nondenaturing gels.*
3. Samples are *not* boiled before application to the gel, since this would lead to thermal denaturation and potential aggregation.
4. During the run, the voltage must be kept low, so that sample heating in the gel is minimized. The voltage recommended above are usually safe for nondenaturing gels as well.

Note that the absence of SDS in nondenaturing gels means that the mobility of the protein will not depend solely on molecular weight here. For this reason, caution should be exercised in using this method for molecular weight determinations. Detailed protocols for determining the molecular weights of proteins from nondenaturing gel electrophoresis have been presented elsewhere (Hedrick and Smith, 1968; Bollag and Edelstein, 1991).

Gradient Gels

Increased band resolution and resolvable molecular weight ranges can be achieved in gel electrophoresis by using a linear gradient of acrylamide concentration rather than a fixed percentage in the separating gel. The preparation of gradient gels is somewhat more tedious than that of fixed percentage gels, and requires some specialized equipment. A low percentage and a high percentage acrylamide solution are mixed to form a linear gradient by use of a gradient former and a peristaltic pump as illustrated in figure 4.6. To prepare a 5-20 % gradient gel, the following procedure can be followed.

MATERIALS

5% ACRYLAMIDE SOLUTION

Combine 1.67 ml of 30% acrylamide/0.8% bisacrylamide, 2.5 ml of 4X Tris-Cl/SDS, pH 8.8, 5.8 ml distilled water, and 50 μl of 10% ammonium persulfate solution. Degas as described above.

20% ACRYLAMIDE SOLUTION

Combine 6.67 ml of 30% acrylamide/0.8% bisacrylamide, 2.5 ml of 4X Tris-Cl/SDS, pH 8.8, 5.8 ml distilled water, 1.5 g sucrose, and 50 μl of 10% ammonium persulfate solution. Degas as described above.

PROCEDURE

1. Prepare the gel apparatus as described previously.
2. Calibrate the peristaltic pump with distilled water so that it provides a flow rate of 3 ml/min.
3. Connect the gradient maker, peristaltic pump, and gel apparatus together as illustrated in figure 4.6.
4. Add 5 μl of TEMED to both the 5% and 20% acrylamide solutions. Swirl to mix, and rapidly add equal volumes (e.g., 5 ml) of the 5% solution to chamber A of the gradient maker, and of the 20% solution to chamber B. Turn on the magnetic stirrer and open the valve between chambers A and B on the gradient maker. Turn on the peristaltic pump to allow the solutions to flow into the gel apparatus. Note that

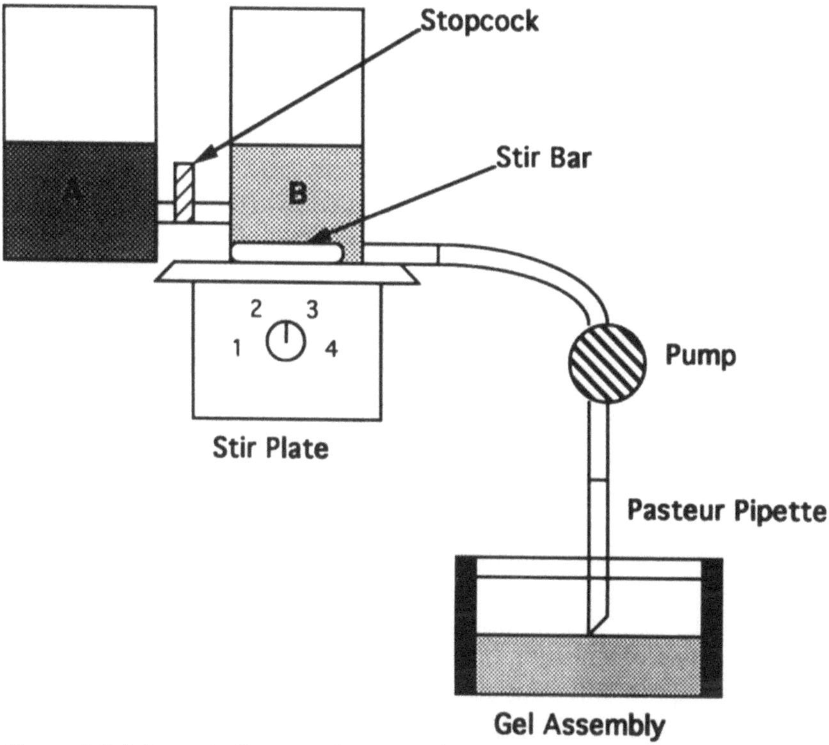

Figure 4.6 Schematic diagram of a set up for preparing gradient gels for electrophoresis. A gradient maker with a stopcock between sample chambers is used to mix the solutions. Solution A contains the lower percentage acrylamide (5% in the example in the text) and solution B contains the higher percentage acrylamide (20% in the example in the text). Solution B also contains sucrose to increase its density. Equal volumes of the two solutions are added to chambers A and B, respectively, and a stir bar is added to chamber B. The apparatus is placed on a magnetic stirrer to provide adequate mixing, and the stopcock between the chambers is opened. As the solutions mix, they flow out of chamber B through tubing and a Pasteur pipette into a gel sandwich assembly to form the gradient gel.

one must work rapidly to prevent the gels from polymerizing before entering the gel apparatus.

5. When the gel is completely poured, layer isobutanol on top of the gel, and proceed from here as described above for denaturing gels.
6. Wash the gradient maker and tubing with warm water to prevent build up of polymerized acrylamide.

Some Comments on SDS-PAGE

The use of SDS-PAGE is very widespread due to the relative simplicity of the technique. Nevertheless, some problems can be encountered with certain samples. Band distortions can occur, for example, if the current used to run the gel is too high, causing excessive heating in the gel. Reduced currents, running the gel in a cold room, or cooling the gel with a circulating water bath can overcome these problems. High concentrations of salts, and some detergents can lead to anomalous staining patterns on gels, that can obscure protein bands. In some cases, proteins will fail to enter the gel, and one will only observe staining near the bottom of the wells in the stacking gel. This is an indication that the protein has precipitated before entering the gel, or that there was significant particulate matter in the sample. Both of these problems can be eliminated by centrifuging the samples just prior to loading.

With many membrane proteins, bands will appear diffuse, or one will observe extensive smearing out of the bands. As discussed above, this problem is often related to aggregates that form during thermal denaturation of these hydrophobic proteins. It is generally better, therefore, not to boil samples of membrane proteins, but rather to incubate them in sample buffer at 37° C for 30 min or more (see above). More complete troubling shooting guides for protein gel electrophoresis have been presented by Hames and Rickwood (1990) and by Andrews (1986). Additional information can be obtained from the technical representatives of the suppliers of electrophoretic equipment (e.g., Bio-Rad, Pharmacia) as well.

A minor problem that is encountered when loading gels is that one is trying to place sample into a clear well that is surrounded by clear gel. More than one researcher has accidently missed the well, because it was hard to see, and loaded the sample into the gel instead. To avoid this problem, my laboratory dissolves a few grains of Bromphenol Blue in 2 ml of distilled water and adds one drop (ca. 50 μl) of this to the stacking gel solution before polymerization. This gives a light blue color to the stacking gel, making it easy to see the clear wells against this background. The inclusion of this small amount of dye does not affect the efficiency of the gel in any way.

The casting of gels, especially gradient gels, is somewhat time consuming and tedious. Some workers prefer to make up a stock of 10 or more gels at one time, and store them at 4° C for later use. Apparatus for the simultaneous pouring of multiple gels are available from most suppliers, and come with detailed instructions for their use, and for storage of gels. For short term storage, gels can be kept for about a week at 4°C in 1X Tris-Cl/SDS, pH 8.8 buffer. An alternative to casting one's own gels is to purchase precast gels. These are available for both denaturing and nondenaturing PAGE, in both single percentage acrylamide and gradient gel formats. Precast gels can save the researcher considerable time and effort, but are more expensive than gels cast in house. Precast gels are available from Bio-Rad, Schleicher & Schuell, Novex, and many other suppliers.

Isoelectric Focusing

Denaturing SDS-PAGE provides us a means of determining protein purity in terms of molecular weight. However, there is always the danger that more than one component, of similar molecular weight, will co-migrate on an SDS-gel. For example, deamidated forms (i.e., forms of the protein in which terminal amino acid residues have been lost) of a larger protein can be present along with the full length protein in solution. The loss of a few residues may not affect the mobility on the gel to a great enough extent for resolution of these species, but it could significantly affect biological activity. How can we test for this type of contaminant? If the species differ in charge at a given pH we can separate them by isoelectric focusing (IEF; Hames and Rickwood, 1990). To understand the basis of this technique we must introduce the concept of isoelectric point.

Isoelectric point = pH at which a molecule has a net charge of zero and will therefore cease to migrate in an electric field.

Consider the amino acid lysine. There are three titratable groups on this molecule:

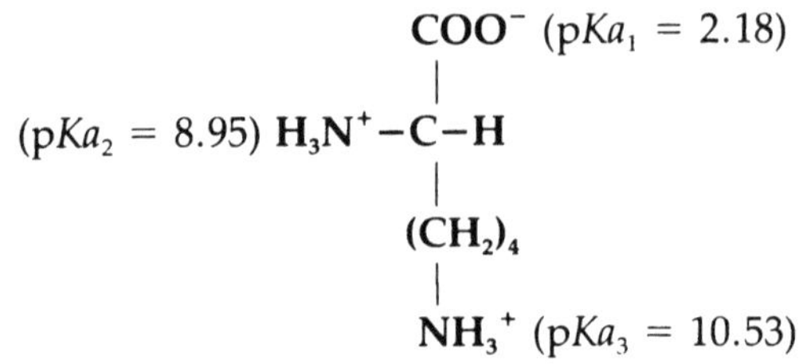

pH titration of this amino acid would lead to the following set of equilibria:

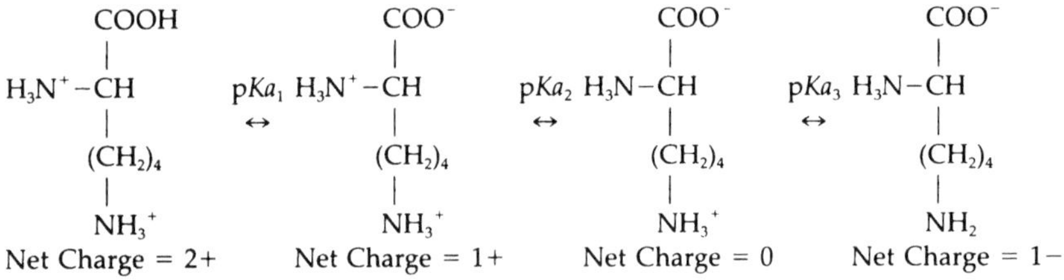

The isoelectric point (pI) is given by the average of the pK_a values involved in the formation of the electrically neutral species. In this case:

$$pI = (pK_{a2} + pK_{a3}) / 2 = (8.95 + 10.53) / 2 = \underline{\underline{9.74}}$$

Now consider the tripeptide:

Lys - Phe - Asp

What is the pI of this peptide?

C-terminal pK_a = 2.2
Lys side chain pK_a = 10.53
Asp side chain pK_a = 4.0
N-terminal pK_a = 9.0

The relevant equilibria are:

$$H_3N^+{-}Lys^+{-}Phe{-}Asp{-}COOH \leftrightarrow H_3N^+{-}Lys^+{-}Phe{-}Asp{-}COO^- \leftrightarrow$$
$$H_3N^+{-}Lys^+{-}Phe{-}Asp^-{-}COO^- \leftrightarrow H_2N{-}Lys^+{-}Phe{-}Asp^-{-}COO^- \leftrightarrow$$
$$H_2N{-}Lys{-}Phe{-}Asp^-{-}COO^-$$

Therefore $pI = (4.0 + 9.0) / 2 = \underline{\underline{6.5}}$

Exactly the same treatment applies to proteins, although it would be tedious to calculate the pI for any reasonable size protein (nowadays there are computer programs that will do this for you). Suppose now that we had in solution two proteins both of molecular weight 35 kDa, one with a pI of 6.0 and the other with a pI of 8.2. If we were to apply SDS-PAGE to this sample we would observe a single band at 35 kDa, and might incorrectly assume that our sample was pure. Instead, suppose we placed our sample in a pH gradient (say from pH 9.5 to 3.5), and applied an electric field across this gradient (cathode at the high pH limit, and anode at the low pH limit). The proteins would migrate towards the anode, under the influence of the electric field, until they reached the pH at which their net charge was zero (their isoelectric points). In this way, the sample could be resolved into its two component proteins, since one would cease migrating at pH 6, and the other would cease migrating at pH 8.2. This is the basis for the technique of isoelectric focusing.

A convenient way to perform IEF is to form the pH gradient within a polyacryamide gel, and run the electrophoresis as described for denaturing SDS-PAGE. Below we shall detail a method for performing IEF in a mini-slab gel format (PAGE-IEF), using the same apparatus as used for SDS-PAGE above. IEF can also be run in tube gels, in granulated beds, and in solution gradients. These alternative techniques are more fully described by Andrews (1986).

MATERIALS

1. Gel Apparatus (see above)
2. 30% acrylamide/0.8% bisacrylamide solution (see above)
3. 2X IEF sample buffer
4. Catholyte Solution (20 mM sodium hydroxide)
5. Anolyte Solution (10 mM phosphoric acid)
6. Ampholyte Solution (from Bio-Rad, Pharmacia, etc.)
7. Urea (Ultrapure grade)
8. 10% ammonium persulfate
9. TEMED
10. Bromophenol Blue
11. 2-mercaptoethanol
12. 1% ampholyte, 5% sucrose solution
13. IEF Fixing Solution.

AMPHOLYTE SOLUTIONS

These can be purchased from a variety of sources, including Bio-Rad and Pharmacia. These companies offer solutions of ampholytes in a number of pH ranges. The more narrow the ampholyte pH range, the greater resolution one can achieve within that range. In the example below, we shall use a broad pH range (3.5–10), which is what one might use in the absence of any prior knowledge of the likely pI for one's sample.

2X IEF SAMPLE BUFFER

Dissolve 4.8 g of urea, 240 μl of ampholyte solution, 1 ml of 20% Trition X-100, 100 μl 2-mercaptoethanol, and 1 mg of Bromophenol Blue in distilled water. Bring the volume to 10 ml with distilled water. Divide into 0.5 or 1 ml aliquots and store at $-20°$ C.

IEF FIXING SOLUTION

Dissolve 29 g of TCA (trichloroacetic acid) and 8.5 g of sulfosalicylic acid in 100 ml of distilled water. When dissolved add more distilled water to a total volume of 250 ml. Store at room temperature.

POURING THE GEL

1. Assemble the mini-gel apparatus as above.
2. Prepare the gel by dissolving 5 g of urea in 4.5 ml of distilled water, 240 μl of ampholyte solution, and 1.67 ml of 30% acrylamide/0.8% bisacrylamide solution. Mix by gently swirling, and degas as described above.
3. Add 20 μl of 10% ammonium persulfate and 16 μl of TEMED to the gel solution. Swirl to mix, and pour this between the glass plates of the gel assembly to the top of the front glass plate. Insert the gel comb as described above, and let the gel polymerize (30 min to 1 h). Note that no stacking gel is used here.
4. When the gel is polymerized, remove the comb and place gel assembly into the electrophoresis tank.
5. Remove any unpolymerized material from the wells.

SAMPLE PREPARATION AND LOADING

Samples must be free of high concentrations of salt or ionic detergents for IEF. If present, these solution components should be removed by dialysis, gel filtration, or TCA precipitation (see Chapter 2).

6. Mix the sample (10–30 μg) with an equal volume of 2X IEF sample buffer. Incubate the sample for at least 5 min at room temperature, and centrifuge to remove any particulate matter.
7. Apply the sample to the bottom of the wells as described above. Fill the space above the samples in the wells with 1% ampholyte, 5% sucrose solution. This will help protect the proteins from the harsh pH conditions of the catholyte solution in the upper buffer chamber.

RUNNING THE GEL

8. Fill the upper buffer chamber with catholyte solution (20 mM sodium hydroxide) and fill the lower buffer chamber with anolyte solution (10 mM phosphoric acid). Note: The catholyte and anolyte solutions should be prepared fresh from 1 M stock solutions on the day of the IEF run.
9. Attach electrodes to the leads, and run the gel at 150 V (constant current) for the first 20–30 min, and then 200 V (constant current) for another 2.5 h.
10. When the run is completed, turn off the voltage, remove the electrodes, and disassemble the gel apparatus as described above. Place the gel in IEF fixing solution and fix for 30–60 min at room temperature with constant agitation (i.e., on a rotary rocker).
11. After fixing, rinse the gel by soaking in destaining solution (methanol/water, 45/55) for 10 min. Protein bands in IEF gels can be visualized by any of the methods previously described for SDS-PAGE (e.g., Coomassie Blue, or silver staining).

Determination of pH Gradient and Sample pI

The pH at different points in the gel can be determined by cutting a strip of the gel from one of the ends, and slicing this into 1 cm sections from top to bottom. Each section is placed in 1 ml of 10 mM KCl for 30 min, and then the pH of the solution is read with a pH meter (Bollag and Edelstein, 1991).

A much better way to determine the pH gradient in the gel is to run a lane containing a set of proteins of well-defined pI along with the sample proteins. IEF standards are available from most of the major suppliers of electrophoretic equipment. The pI values for proteins that are commonly used as IEF standards are presented in table 4.3. The pI of one's target protein can be estimated by where it runs in the gel relative to the IEF standards (see figure 4.7).

Table 4.3 Proteins that are useful as markers for isoelectric focusing studies. Data taken from the Sigma Chemical Company catalog, St. Louis, MO, page 1663.

Protein	pI (25° C)
Amyloglucosidase (from *A. niger*)	3.6
Glucose oxidase (from *A. niger*)	4.2
Soybean Trypsin Inhibitor	4.6
β-Lactoglobulin A (from bovine milk)	5.1
Carbonic Anhydrase II (from bovine erythrocytes)	5.4 and 5.9
Carbonic Anhydrase I (from human erythrocytes)	6.6
Myoglobin (from equine heart)	6.8 and 7.2
L-Lactic Dehydrogenase (from rabbit muscle)	8.6
Trypsinogen (from bovine pancreas)	9.3
Lysosyme (from hen egg)	10.0

Two-Dimensional Gel Electrophoresis

Two-dimensional gel electrophoresis combines the resolving power of SDS-PAGE with isoelectric focusing. One typically runs an IEF gel as the first dimension, then subjects the target protein lane from that gel to standard SDS-PAGE, to resolve species by molecular weight, in the second dimension. Thus, in one gel, one can discriminate among proteins in a sample both on the basis of molecular weight and isoelectric point (see Andrews (1986) for an excellent review of two-dimensional gel electrophoresis).

PROCEDURE

1. Assemble gel apparatus as described above.
2. For the first dimension, run an IEF gel as discussed in this chapter.
3. While the IEF gel is running, assemble the apparatus and pour a separating and stacking gel for SDS-PAGE as described earlier in this chapter. Pour the stacking gel to within 0.5 cm of the top of the front glass plate. Instead of inserting a comb, cover the stacking gel with distilled water.
4. When the IEF gel is done, cut a 0.5 cm strip from the lane containing the target protein sample (here, and in subsequent steps, handle the strip with gloved hands). Notch one corner of the strip to identify the direction of the pH gradient, and place the strip in a plastic weighing boat. Cover the strip with equilibration buffer (see below) and equilibrate for 30 min on a rotary rocker.

EQUILIBRATION BUFFER

Combine 5 ml of 2-mercaptoethanol, 6.25 ml 1 M Tris-Cl, pH 6.8, 23 ml of 10% SDS, and 10 ml of glycerol. Bring the volume to 100 ml with distilled water.

5. Aspirate away the water from the top of the stacking gel, and replace it with 1X SDS/electrophoresis buffer (see above). Carefully pick up the gel strip and lower one end into the space above the stacking gel, and then lower the other end so that the strip fills the space between the stacking gel and the top of the front plate without any trapped air bubbles. Be careful not to distort the strip during this process, and be sure to note the direction, in terms of the pH gradient, in which you loaded the strip.
6. Run the SDS-PAGE as described earlier in this chapter, and stain the resulting gel with one of the standard methods described above.

Figure 4.8 schematically illustrates the expected results of a two-dimensional gel run by the protocol described here for a hypothetical protein solution of 4 proteins of molecular weights 60 and 35 kDa, and pI's of 6.0 and 8.5. Alternative protocols for two-dimensional gels are described by Hames and Rickwood (1990).

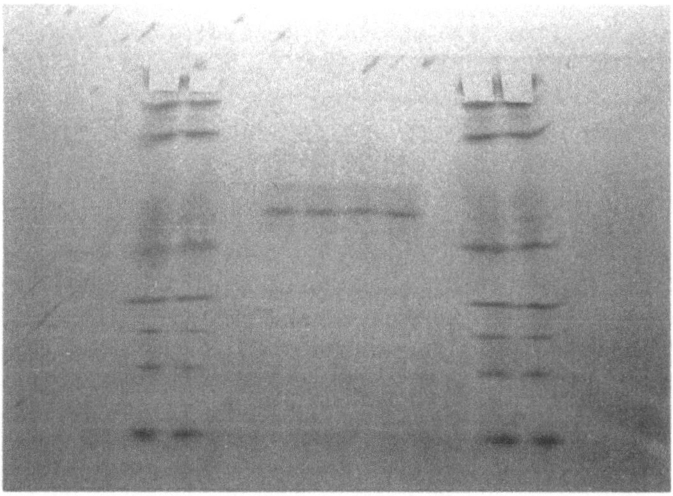

Figure 4.7 IEF gel of samples of horse heart myoglobin showing two or more isoforms of the protein. The end lanes on each side of the gel contain protein standards of known pI (from Pharmacia). Data provided by Jason Felsch, Dept. of Biochemistry, The University of Chicago.

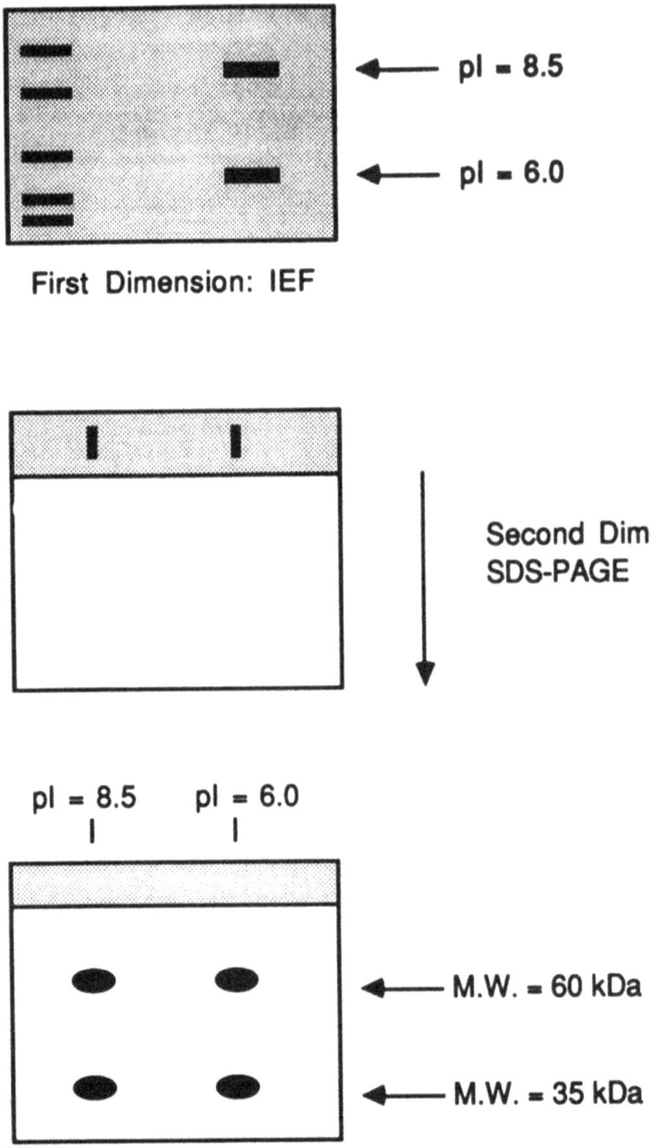

Figure 4.8 Schematic representation of a two-dimensional gel experiment. Here four proteins are present in the sample with the following molecular weights and isoelectric points: MW = 60 kDa, pI = 8.5; MW = 60 kDa, pI = 6.0; MW = 35 kDa, pI = 8.5; and MW = 35 kDa, pI = 6.0. The proteins are first separated by IEF. The lane containing the sample is cut out of the IEF gel in a 3 cm strip and laid atop a resolving gel for SDS-PAGE. The samples are electrophoresced into the resolving gel and separated by molecular weight in the second dimension.

Capillary Electrophoresis

The idea of separating molecules electrophoretically in narrow bore capillaries has been around since the late 1960's. It was not until very recently, however, that these ideas have been reduced to practice. Today, several commercial units for capillary electrophoresis (CE) are available, and their applications to protein separation have been documented.

CE achieves molecular separations on the same basis as conventional electrophoretic methods, but does so within the environment of a narrow capillary tube. The main advantages of CE are that very small (nanoliter) volumes of sample are required, and because of the capillary formate, separation and detection can be performed rapidly, thus greatly increasing sample throughput relative to gel electrophoresis.

The instrumentation used for CE is in some ways a hybrid between electrophoretic and HPLC designs. Figure 4.9 illustrates a schematic for a basic CE instrument. The main components of a commercial CE unit are an electrophoretic chamber containing a capillary, two buffer reservoirs connected to opposite ends of the capillary, a dual high voltage power supply, and a detector. The molecules are separated as they are conducted, at variable rates, through the electrolyte-filled capillary tube, by the action of an applied electric field (typically 200–400 V/cm). The two forms of CE that have been most widely applied to the separation of proteins and peptides are capillary zone electrophoresis (CZE), which separates molecules on the basis of differences in mass to charge ratios, and isoelectric focusing (CE-IEF), which separates molecules on the basis of differences in isoelectric point (*vide supra*).

Samples are introduced into the capillary unit in a number of ways. Electrophoretic loading is accomplished by applying a high voltage while one end of the capillary is in contact with the sample solution. Ionic species in the sample will enter the capillary at rates depending on their electrophoretic mobilities. Another common loading strategy is displacement loading, in which a positive pressure of inert gas is applied to the sample container, or a vacuum is applied to the outlet end of the capillary to force the flow of sample into the capillary tube. Gravity loading is also sometimes used, in which the capillary tube is filled with sample by displacing the relative heights of the sample reservoir and capillary tube.

The capillary tubes used in commercial CE units are invariably made of fused silica, because of the low UV absorbance of this material. Fused silica is, however, rather fragile, and so the tubes are usually coated with a polymer, such as polyimide, to reduce the chances of damage to the

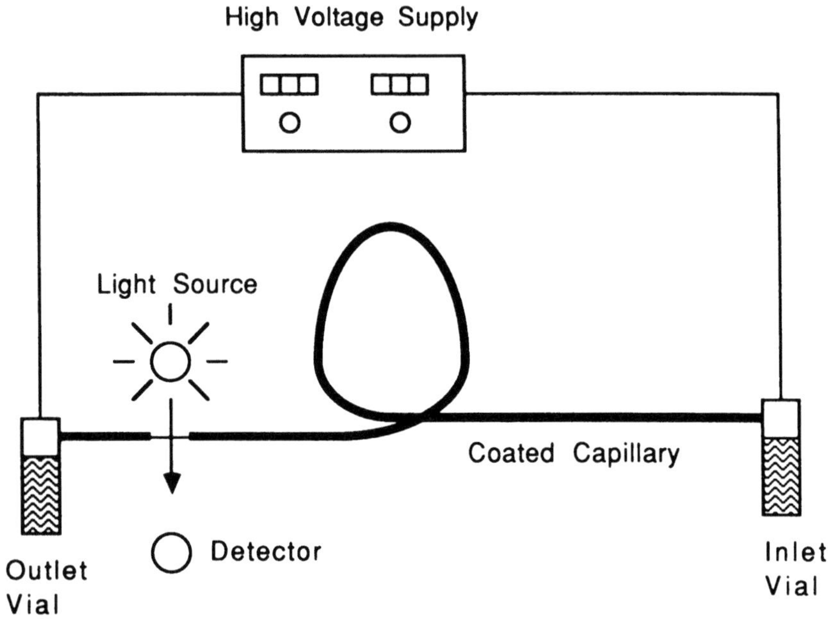

Figure 4.9 Schematic diagram of a typical commercial capillary zone electrophoresis apparatus as described in the text.

tube. A small section of the tube must be left uncoated to serve as a detection window, since the polymer coating interferes with UV detection.

Although a variety of detection methods have been employed for CE, most commercial units use UV absorbance (200 or 280 nm) as the detection method. Detection thus depends on Beer's law, and the narrow cross sections of the capillary tubes used in CE present a detection problem since they result in very small optical pathlengths. This problem could be overcome by increasing the inner diameter of the capillary, but larger-bore capillaries lead to reduced band resolution due to less effective dissipation of thermal gradients caused by the current flow during electrophoresis. Thus, the size of the capillary utilized is a compromise between these two factors. The best compromise appears to occur with tubes of inner diameters between 25 and 50 μm.

The use of fused silica for the capillary tubes presents an additional potential problem. At pH 3 or higher, the silica will have a negative surface charge that will attract cations from the buffer, thus creating an electrical double layer at the inner surface of the capillary. During electrophoresis at neutral or alkaline pH, cations within this double layer will migrate, causing the bulk fluid to move towards the negative electrode in a process known as electroendosmotic flow. Proteins in solution will adsorb to the ionized sites on the capillary, causing variations in the rate of electroendosmotic flow, and thus difficulties in reproducing migration times and integrated peak areas. To overcome this problem, polymer-coated capillaries are used. The polymer coating significantly reduces both protein adsorption and electroendosmotic flow, thus providing dramatically improved resolution for protein samples.

Applications of CE for Proteins

As stated above, the two most commonly used forms of CE for proteins are capillary zone electrophoresis (CZE) and isoelectric focusing (CE-IEF). In CZE molecules are separated on the basis of differences in their charge to mass ratios. For small peptides, the total charge and molecular weight can be calculated from the amino acid sequence, and thus one can predict the relative mobilities of different peptides in CZE. This method has been used to separate small peptides, for example, from proteolytic digestions of proteins (see Chapter 7 for a discussion of proteolytic digestion). Figure 4.10 illustrates the type of resolution obtained with CZE for small peptides. Here nine peptides are well resolved at low pH using a coated capillary tube. The entire electrophoretic run took less than 15 min.

IEF is performed in coated capillaries using the same basic principles as described earlier for conventional IEF. Samples are premixed with ampholytes before introducing them into the capillary by displacement loading. A voltage of 6–8 kV is then applied for a few minutes, during which a pH gradient develops along the length of the capillary and the proteins migrate within this gradient to their isoelectric points. After this, the cathode solution (NaOH) is supplemented with NaCl. The chloride ions enter the capillary and cause the pH at the outlet end to drop; this results in migration of the focused proteins towards the detector with the most basic proteins eluting first. Because of the UV absorption properties of the ampholytes used in IEF, sample detection must be done at 280 nm.

Whether one uses CZE or CE-IEF, the chief advantage of capillary electrophoresis is the rapid sample throughput that can be achieved. If one needs to screen large numbers of samples on a routine basis, CE may be the technique of choice over conventional gel systems. Commercial CE units are now available from Bio-Rad, Beckman, and several other

Peptide	MW
1. Bradykinin	1060
2. Angiotensin II	1046
3. α-Melanocyte stimulating hormone	1665
4. Thyrotopin releasing hormone	362
5. Luteinizing hormone releasing hormone	1182
6. [2 - 5] leucine enkephalin	392
7. Bombesin	1620
8. Methionine enkephalin	574
9. Oxytocin	1007

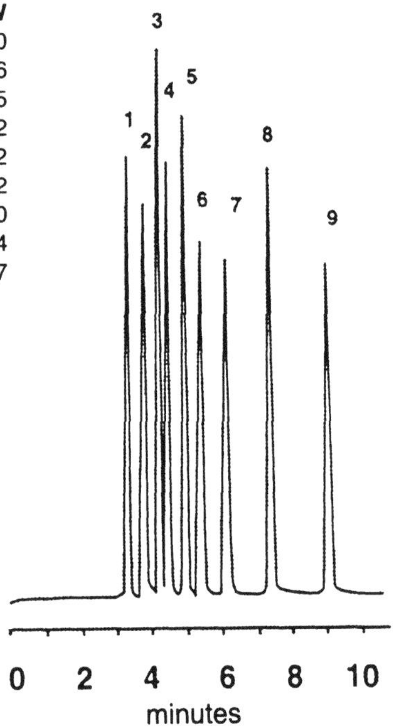

Capillary: 20 cm × 25 Mu, coated

Butter: 0.1 M sodium phosphate, pH 2.5

Load conditions: 8 kV, 8 seconds

Run conditions: 8 kV, constant voltage, ⊕→⊖ polarity

Detection: UV, 200 nm, 0.02 AUFS

Figure 4.10 Results of a typical CZE size exclusion experiment for the separation of nine small peptides: (1) Bradykinin; (2) Angiotensin II; (3) α-Melanocyte stimulating hormone; (4) Thyrotropin releasing hormone; (5) Luteinizing hormone releasing hormone; (6) [2–5] leucine enkephalin; (7) Bombesin; (8) Methionine enkephalin; (9) Oxytocin. Samples were loaded for 8 s using 8 kV onto a 20 cm × 25 μm coated capillary in 0.1 M sodium phosphate buffer, pH 2.5. Electrophoresis was carried out at 8 kV constant voltage in the same buffer. Peptide peaks were detected by absorbance at 200 nm. Data provided by Bio-Rad Laboratories, Life Science Group, Analytical Systems Division, Hercules, CA (with permission).

manufacturers. An excellent review of the theory behind CZE and its applications to separation of peptides and proteins is provided by Schwartz et al. (1992).

CHROMATOGRAPHIC METHODS

Chromatographic methods are so widely used in analytical sciences that entire texts and series have been devoted to this subject. Here we shall briefly describe some of the more commonly used methods for assessing the purity of proteins and peptides. While ambient pressure chromatography is widely used in protein purification, these methods have largely been replaced by high performance liquid chromatography (HPLC), at elevated pressures, for analytical purposes. HPLC provides significantly enhanced resolution and turnaround time over ambient pressure systems. Three forms of HPLC are commonly employed in protein purity assessment: size exclusion, ion exchange, and reverse phase HPLC (see Hancock (1984) for a review of these methods).

Size Exclusion HPLC

The principles behind size exclusion HPLC are the same as those discussed for ambient pressure size exclusion chromatography in Chapter 2. The major difference here is that the stationary phase materials must be capable of withstanding the high pressures used in HPLC without physical deformation. The experimental observables that must be measured in order to describe the migration of a protein on a size exclusion column are the void volume (V_0) of the column, i.e., the interstitial liquid volume between particles of the stationary phase matrix, and the elution volume of the protein (V_e). The void volume is typically determined by loading a solute with a molecular weight far in excess of the exclusion limit of the stationary phase, and observing the volume of mobile phase that is required to elute this solute from the column. Dextran Blue is commonly used for this purpose (available from Sigma), since the bright blue color of this polymer makes for easy observation of elution. Once V_0 and V_e have been established, one can calculate the relative elution volume (*REV*) for the protein, which reflects the degree to which the protein is retarded on the column:

$$REV = V_e/V_0$$

Were one to load several proteins of known molecular weight onto a size exclusion column and compute their *REV* values, one would find that *REV* varies linearly with the log of molecular weight for the proteins, as illustrated in figure 4.11. Thus, by loading a sample of a putatively pure protein onto a size exclusion column, one can not only assess the purity of the sample (by calculation of the number of peaks and their relative integrated areas), but also estimate the molecular weight of the protein by reference to the *REV* values of molecular weight standards. Standard kits of proteins with well-defined molecular weights are available commercially for this purpose.

Size exclusion HPLC can be run under native or denaturing conditions, as well as under reducing or non-reducing conditions, depending on the mobile phase used. This provides the protein scientist with considerable versatility in characterizing a target protein by HPLC. When run under native conditions, the apparent molecular weight observed can be different from the true molecular weight of the protein because of electrostatic or hydrophobic interactions between the protein and the stationary phase. These types of interactions are minimized by appropriate choice of buffer and by avoiding very low ionic strength conditions. Buffers containing 0.15 M NaCl, for example, usually work quite well. Of course, optimal buffer conditions will depend on the target protein and (to a lesser extent) the type of column used. Note that under native conditions, the molecular weight determined by size exclusion HPLC reflects the molecular weight of the native molecular unit of the protein. This can be different from the denatured molecular weight for multi-subunit proteins, and for proteins which normally do not exist as monomers. It should also be noted that while size exclusion HPLC is widely used to estimate the molecular weights of globular proteins, the elution profile of a protein will depend not only on its molecular weight, but also on its overall shape (i.e., hydrodynamic radius) as discussed in Hancock (1984).

Comparison of the elution profiles from size exclusion HPLC run under reducing and non-reducing conditions can provide one with information on the presence of disulfide bonds within a protein. When coupled with proteolytic digestions, this method can further be used to determine the arrangement of disulfide bonds within a protein. The use of size exclusion HPLC for this purpose is discussed in Chapter 7.

Ion Exchange HPLC

Because of their charged character, proteins will interact electrostatically with the charged surfaces of certain stationary phases. These electro-

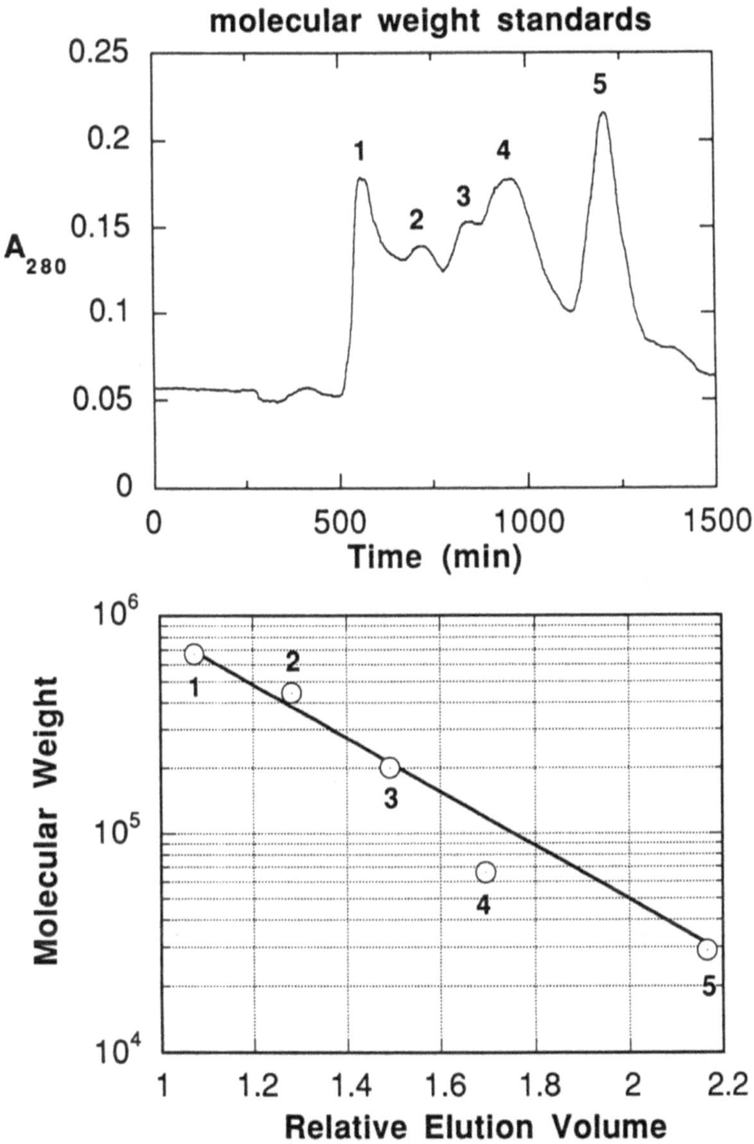

Figure 4.11 Elution profile of five proteins of known molecular weight from a size exclusion column. The top panel shows the absorbance at 280 nm as a function of time during the elution of samples from the column. The bottom panel plots the relative elution volume for each protein as a function of the log of its molecular weight (a Ferguson plot). Data provided by Aldona Balciunas and Martin Horvath, Department of Biochemistry and Molecular Biology, The University of Chicago.

static interactions can then be disrupted by increasing the ionic strength of the mobile phase. The exact ionic strength required to elute a particular protein will depend on the strength of interaction between the protein and the mobile phase. Typically, what is done is to load the protein onto the column under low ionic strength conditions, where interactions are maximized, and then elute the protein with a linear gradient of salt concentration. Thus, ion exchange chromatography can be used to separate proteins on the basis of differences in ionic charge, and has therefore been widely used in protein purification and analysis.

Two basic types of ion exchange resins are commercially available: anion exchange resins, and cation exchange resins. The choice of which to use for a specific protein will depend on the isoelectric point of that protein. Anion exchange HPLC is used for proteins with isoelectric points below pH 7.0, while cation exchange HPLC is used for proteins with isoelectric points above pH 7.0. The Mono Q and Mono S columns from Pharmacia are particularly widely used for ion exchange HPLC. My own laboratory has used these columns extensively in purifying a variety of proteins with very satisfactory results.

In order for the proteins to interact strongly, and hence stick well to the stationary phase, they should be loaded onto the column at an ionic strength less than or equal to the starting point of the elution salt gradient. The column must be well equilibrated with starting buffer before loading the sample. If one wishes to run ion exchange under denaturing conditions, a non-ionic denaturant such as urea should be used (see Chapter 10 for a discussion of denaturants and their effects on proteins). Likewise, if one uses a detergent to solubilize a membrane protein, a non-ionic detergent, such as Tween-20 (Pierce) or dodecyl-β-maltoside (Anatrace) should be used for ion exchange HPLC work. An example of the type of resolving power obtained with ion exchange HPLC is illustrated in figure 4.12.

Reverse Phase HPLC

In reverse phase HPLC proteins are separated on the basis of differences in their strength of hydrophobic interaction with the hydrophobic surface of the stationary phase material. Proteins are usually eluted from these columns by gradually increasing the amount of a non-polar organic solvent in the mobile phase. Proteins elute from reverse phase columns in the reverse order of their strengths of interaction with the stationary phase, i.e., those proteins that interact weakly with the stationary phase will elute at lower organic solvent concentrations, whereas those that

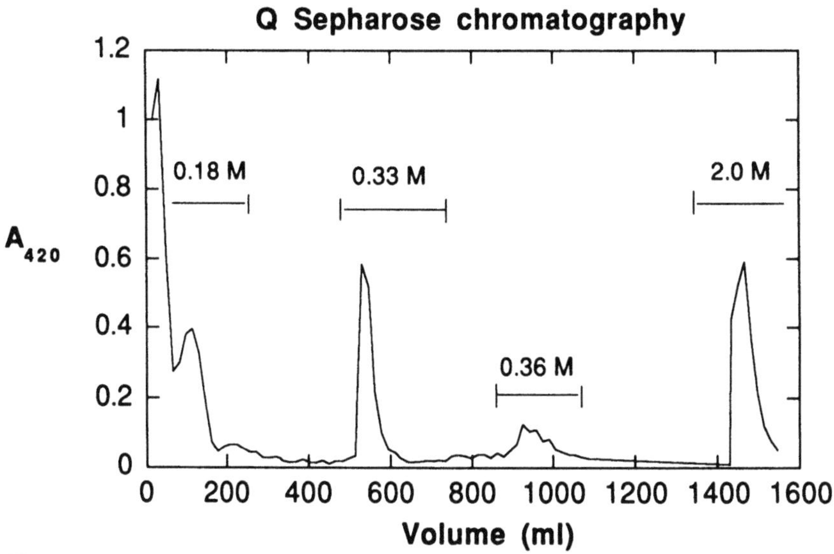

Figure 4.12 Elution profile of heme-containing proteins from an ion exchange column. Here proteins from the periplasmic membranes of cells of *Paracoccus denitrificans* were loaded onto a Q sepharose column at low ionic strength, and eluted by increasing the ionic strength of the mobile phase in steps (as indicated on the figure). The absorbance at 420 nm was monitored as a function of elution volume in order to detect heme-containing proteins that absorb light at this wavelength. Data provided by Aldona Balciunas and Martin Horvath, Department of Biochemistry and Molecular Biology, The University of Chicago.

adsorb strongly will require higher concentrations of organic solvent to elute. The elution pattern for a particular protein will vary with the hydrophobic character of the stationary phase, the organic solvent used in the mobile phase, and the gradient of organic solvent used.

The stationary phases used for reverse phase HPLC are almost always alkyl silanes. The hydrophobicity of the column will depend on the alkyl chain length, and the degree of silica group protonation. Sample elution from these columns is most often accomplished by running a gradient from high percent aqueous solvent/low percent organic solvent to low percent aqueous solvent/high percent organic solvent. Both mobile phase solutions usually contain a relatively strong acid, such as trifluoroacetic acid (TFA). The inclusion of TFA in the mobile phase helps in keeping proteins soluble, and also helps to keep the amount of deprotonated silanols on the surface of the stationary phase to a minimum. A commonly used organic solvent for reverse phase HPLC is acetonitrile. A good general mobile phase gradient for proteins and peptides is as follows:

Solvent A: 0.1% TFA in distilled water.

Solvent B: 0.085% TFA, 70% acetonitrile in distilled water.

Load the sample, and start the gradient in 100% solvent A. Elute with a linear gradient from 100% solvent A to 100% solvent B.

Note that the amount of TFA is slightly different in the two solvent systems. This is because of differences in the deep ultraviolent molar absorbance of TFA in water and acetonitrile. The small change in percentage of TFA helps to reduce baseline drifts when using far ultraviolet detection (*vide infra*).

One important caveat that must be kept in mind is that the conditions commonly used for reverse phase HPLC (i.e., low pH, high percent organic solvents) are inherently denaturing to most proteins. While HPLC is a very powerful analytical tool for detecting proteins, the proteins eluted from such columns should not, in general, be considered to be properly folded. It is very common to observe significant diminutions of biological activity for proteins after reverse phase HPLC. Therefore, **unless there is strong evidence to the contrary, one should assume that proteins eluted from a reverse phase column are denatured.** In some cases, the denatured proteins obtained from a reverse phase column can be refolded into their native conformations. A description of methods for refolding proteins is presented in Chapter 10.

Sample Preparation and Detection

Detection of proteins eluting from HPLC columns can be performed in a number of ways, including radioactive detection of isotopically labeled proteins. Most commonly, however, proteins and peptides are detected by ultraviolet absorbance or fluorescence. Absorbance at 214 or 220 nm is most commonly used for detection of proteins and peptides. At these wavelengths detection is based on the absorbance of the amide bonds within all peptides. Because of the high copy number of amide bonds in proteins and peptides, detection at these wavelengths is quite sensitive, and optical density per mass of protein is fairly constant from protein to protein. A major drawback of detection at these wavelengths is that many non-protein components of buffer systems also absorb light in this spectral region. An alternative method is to detect absorbance at 280 nm, where tyrosine and tryptophan residues within proteins absorb strongly. Detection at this wavelength is more free of interferences from non-protein sources, but the sensitivity is lower here, because of the lower content of these amino acids relative to amide bonds in proteins. Another problem with detection at 280 nm is that only proteins and peptides that contain the aromatic amino acids tyrosine and tryptophan can be detected. This is usually not a problem for detection of globular proteins, but can be a drawback for detecting smaller peptides. A third common detection method for proteins is the use of fluorescence detection. Here one excites the elutant at 280 nm, where tyrosine and tryptophan absorb, and monitors fluorescence at wavelengths between 310 and 360 nm. The fluorescence from these aromatic residues is quite strong, increasing the detection sensitivity about 100-fold over 280 nm absorbance. However, this method is again limited to those proteins and peptides that contain tyrosine and/or tryptophan residues. An additional drawback of fluorescence detection is that the instrumentation is somewhat more costly than absorbance detectors.

Regardless of what type of column is used for HPLC analysis, it is important that the protein sample be completely solubilized before loading. Particulate matter in the sample can clog columns and affect the elution patterns observed. Prior to loading onto the column, samples should be passed through a 0.22 μ filter to remove any insoluble materials. Likewise, solvents used in the mobile phases for HPLC must be of high quality and free of any particulate matter. Dissolved oxygen in the mobile phase can lead to cavitation problems at the high pressures used for HPLC. It is thus a common practice to sparge mobile phase solutions with nitrogen to remove dissolved gases. HPLC grade solvents are available from most scientific supply houses, and detailed instructions for solvent and sample preparation are provided by most of the commercial HPLC instrument manufacturers (Waters, Pharmacia, Bio-Rad, Hewlett-Packard, etc.). An excellent text devoted to HPLC of proteins and other macromolecules has recently been published (Oliver, 1989), and provides considerable detail on these methods.

References

Andrews, A. T. (1986) *Electrophoresis: Theory, Techniques and Biochemical and Clinical Applications*, 2d ed., Oxford University Press, Oxford.

Ausubel, F. M.; Brent, R.; Kingston, R. E.; Moore, D. D.; Seidman, J. G.; Smith, J. A.; and Struhl, K. (1989) *Short Protocols in Molecular Biology*, Wiley, New York.

Bollag, D. M., and Edelstein, S. J. (1991) *Protein Methods*, Wiley, New York.

Cooper, T. (1977) *The Tools of Biochemistry*, Wiley, New York.

Hames, B. D., and Rickwood, D. (1990) *Gel Electrophoresis of Proteins: A Practical Approach*, 2d ed., IRL Press, London.

Hancock, W. S. (1984) *Handbook of HPLC for Separation of Amino Acids, Peptides, and Proteins,* CRC Press, Boca Raton, FL.

Hedrick, J. L., and Smith, A. J. (1968) *Arch. Biochem. Biophys.,* **126,** 155–164.

Kadenbach, B.; Jarausch, J.; Hartmann, R.; and Merle, P. (1983) *Analyt. Biochem.,* **129,** 517–521.

Oliver, R. W. (1989) *HPLC of Macromolecules: A Practical Approach,* IRL Pres, London.

Schwartz, H. E.; Palmieri, R. H.; Nolan, J. A.; and Brown, R. (1992) *Separation of Proteins and Peptides by Capillary Electrophoresis: An Introduction,* Beckman Instruments, Inc., Fullerton, CA.

5

Immunological Methods

The specificity with which antibodies react with their target forms the basis of a number of analytical methods that can be used to confirm the identity of a protein, and to screen samples (e.g., column fractions) for the presence of a specific protein. Two of the most common immuno-chemical methods for protein analysis are Western blots and antibody-capture assays. These shall be described in detail in this chapter. Descriptions of other immuno-chemical methods, such as rocket electrophoresis and immunoprecipitation can be found in several texts that are devoted exclusively to immuno-chemistry (Axelsen, 1975; Bjerrum and Heegaard, 1988; Harlow and Lane, 1988). Before describing these analytical methods, a brief overview of immunology is presented, so that the terms used later in this chapter are clearly defined. A more thorough discussion of antibodies and immunology can be found in the text by Harlow and Lane (1988).

OVERVIEW

To guard against infection by foreign molecules and organisms, the immune systems of higher organisms produce three types of lymphocytes that together are responsible for immune response: B cells, cytotoxic T cells, and helper T cells. B cells secrete antibodies that can bind to foreign proteins and other macromolecules in solution. These cells also carry modified versions of the antibody on their cell surfaces that act as receptors for the foreign macromolecule. T cells also have modified antibodies on their cell surfaces, but these only recognize molecules on the surfaces of foreign cells. It is important to note that each lymphocyte produces only one antibody that recognizes one specific target molecule.

An antibody is defined as a member of a class of proteins, known as immunoglobulins, that recognize and bind a specific molecular target. The terms antibody and immunoglobulin are often used interchangeably, although strictly speaking the term antibody should be reserved for immunoglobulins whose target is well defined. The ability of a molecule to be recognized and bound by an antibody is referred to as the antigenicity of that molecule; a molecule that is the binding target for an antibody is thus termed an antigen. The exact structural unit on an antigen that is recognized by an antibody is known as an epitope.

All antibodies are composed of a single or multiple repeats of a Y-shaped basic structural unit (figure 5.1). This basic unit is itself composed of four protein subunits, two heavy chain subunits, and two light chain subunits. The number of repeats of the basic Y-shaped unit that a particular antibody contains (IgG – 1; IgM – 5; IgA – 2, 4, or 6; IgD – 1; IgE – 1) and its heavy chain subunit type (IgG – γ; IgM – μ; IgA – α; IgD – δ; IgE – ε) define the class of antibody to which the molecule belongs. The IgG class consists of the structurally most simple antibodies, containing only a single Y-shaped unit per molecule. These proteins contain three domains, two forming the arms of the Y, and a third forming the

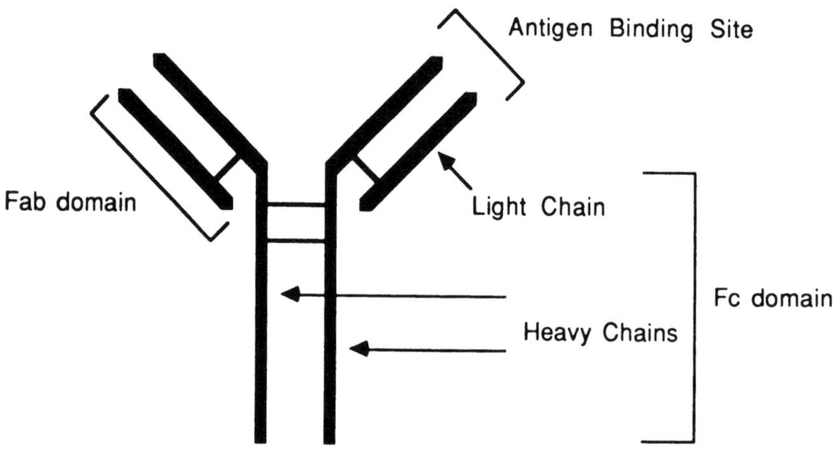

Figure 5.1 Diagram of an immunoglobulin G molecule. The heavy and light chains are covalently linked to each other by a disulfide bond, and the two heavy chains are held together by two disulfide bonds. Each Fab fragment contains a single antigen binding site so that the overall molecule is bidentate (i.e., binds two molecules of antigen per IgG molecule) for antigen.

base of the Y. These domains can be separated and isolated by treatment of the IgG with the proteolytic enzyme papain. The two arms of the Y each contain a binding site for an antigen, and are referred to as Fab fragments (**f**ragments containing the **a**ntigen **b**inding capacity). The domain forming the base of the Y is referred to as the Fc fragment (Harlow and Lane, 1988).

When an animal is infected with a foreign antigen by, for example, injection of a purified target protein, B cell surface antibodies bind to the antigen and are activated to secrete antibody and to proliferate. The ensuing burst of cell division, known as a clonal expansion, greatly increases the number of antigen-specific lymphocytes that can secrete antibodies that recognize the antigen. If blood is drawn from the animal at this point, the serum will be rich in antibodies specific to the target antigen. Several B cells may recognize different epitopes on the antigen and produce antibodies to these epitopes. Likewise, the antibodies to antigens other than the target protein may be present in the serum as well. Thus, antibodies obtained in this fashion are not homogeneous. Since they do not necessarily derive from clonal expansion of a unique B cell, they are referred to as polyclonal antibodies. For many analytical purposes, polyclonal antibodies, while not homogeneous, are of adequate purity.

Antigen-specific B cells can be isolated from immunized animals and fused with a B cell tumor cell, known as a myeloma cell. The resulting hybrid cells can be grown in cell culture, and will continuously secrete antibody. Since these *hybridomas* are derived from a specific B cell, they represent clonal progenies of that cell, and thus secrete a specific, homogeneous antibody population. Antibodies derived from this process are referred to as monoclonal antibodies.

The details of animal immunization and antibody production are outside the scope of this book. The reader interested in greater detail can find an excellent discussion of these methods in the recent text by Harlow and Lane (1988). Most laboratories that use antibodies for analytical purposes do not produce the antibodies themselves. A number of commercial laboratories exist that will produce and supply either polyclonal or monoclonal antibodies raised against the customer's target antigen. It follows, of course, that the quality of these antibodies will depend on the purity of the antigen sample used to immunize the animals for antibody production. In general one should try to provide protein samples in as high a state of purity as possible. Purity criteria for antigens can be obtained from the commercial laboratories that produce antibodies, and are discussed by Harlow and Lane (1988).

In the analytical methods that we shall discuss next, one uses a primary antibody that is specific for the target protein one wishes to analyze for. A labeled secondary antibody is then used to facilitate detection of the primary antibody (and hence the target protein). A secondary antibody is one that recognizes and binds to another immunoglobulin as its antigen. Cross reaction between the primary and secondary antibodies will depend upon the animal species used in their production. For example, if one raises a primary antibody by immunization of mice with one's target protein, one might use a goat anti-mouse secondary antibody. This means that the secondary antibody is produced in goats by immunization of the animal with immunoglobulins derived from mice. The cross reactivity can also be dependent on the class (IgG, etc.) of antibody. Commercially available labeled secondary antibodies usually are raised against IgG immunoglobulins. Thus, one can purchase secondary antibodies such as goat anti-rabbit IgG, goat anti-mouse IgG, and goat anti-human IgG that will bind specifically to IgG class antibodies raised in rabbits, mice, and humans, respectively.

Labeling of the secondary antibody can be accomplished in a number of ways. The important criterion here is that the label provide a convenient means of detection, while not interfering with the antibody-antigen interaction. Secondary antibodies are labeled by iodination with radioactive ^{125}I, by coupling to enzyme systems that can be used with chromogenic substrates for color production, and by coupling with colloidal gold. In the protocols that follow, we shall describe the use of secondary antibodies coupled to alkaline phosphatase, an enzyme-based method that leads to color production. I have chosen to focus on this detection method because of its general utility and convenience. Discussions of other detection methods can be found in several texts (Harlow and Lane, 1988; Andrews, 1986; Bollag and Edelstein, 1991; Bjerrum and Heegard, 1988) and in the product literature from different manufacturers (for example Bio-Rad Laboratories provides a useful handbook on protein blotting; bulletin 1721).

WESTERN BLOTS

Western blotting is an electrophoretic technique that allows one to test the cross reactivity of individual protein bands on an SDS-PAGE or IEF gel, with an antibody raised against a specific antigen. This method is often used to verify the identity of a protein band as the target protein. In Western blotting, proteins are transferred electrophoretically from the gel to a sheet of nitrocellulose, which binds proteins well. The rest of the nitrocellulose is saturated (blocked) with non-antigenic protein, usually

BSA or non-fat dry milk, to prevent non-specific immunoglobulin binding to the nitrocellulose. The nitrocellulose is then treated with a primary antibody raised against the target protein, and next with a labeled secondary antibody that binds to the primary antibody. In this way the presence of the target protein and its migration pattern on the gel can be determined (Towbin et al., 1979).

MATERIALS

TOWBIN'S TRANSFER BUFFER (25 mM TRIS, 190 mM GLYCINE, 20% METHANOL)

Dissolve 2.9 g of Tris and 14.5 g of glycine in 500 ml of distilled water. Add 200 ml of methanol and mix. Bring volume to 1 liter with distilled water. Store at 4° C (Towbin et al., 1979).

BLOCKING SOLUTION

Dissolve 5 g of non-fat dry milk (e.g., Carnation) in 50 ml of TBS (see below). Bring volume to 100 ml with TBS. Store at 4° C. Note this solution degrades within a week or more, and should be prepared fresh. Somewhat longer shelf life can be achieved by adding 0.02% azide as a bacterial growth inhibitor.

10X TBS (TRIS BUFFERED SALINE; 100 mM TRIS, 1.5 M NaCl)

Dissolve 12.11 g of Tris base, 87.66 g NaCl, and 39 ml of 1N HCl in 500 ml of distilled water. Mix well and adjust pH to 7.5 with additional 1N HCl (if necessary). Bring volume to 1 liter with distilled water. To make TBS dilute 50 ml of 10X TBS into 450 ml of distilled water.

10X PBS

Dissolve 2.0 g KCl, 2.0 g KH_2PO_4 (monobasic), 21.6 g $Na_2HPO_4 \cdot 7H_2O$ (dibasic), and 80 g of NaCl in 500 ml of distilled water. Bring volume to 1 liter with distilled water.

PBS/EDTA

Dissolve 0.58 g of EDTA and 10 ml of 10X PBS in 50 ml of distilled water. Bring volume to 100 ml with distilled water.

NBT STOCK

Dissolve 0.5 g of NBT (nitro blue tetrazolium; available from Bio-Rad) in 10 ml of 70% dimethylformamide. Store at 4° C; the solution is stable for a year or more.

BCIP STOCK

Dissolve 0.5 g of BCIP (bromochloroindolyl phosphate, disodium salt; available from Bio-Rad) in 10 ml of 100% dimethylformamide. Store at 4° C; the solution is stable for a year or more.

ALKALINE PHOSPHATASE BUFFER (100 mM TRIS, 100 mM NaCl, 5 mM $MgCl_2$, pH 9.5)

Dissolve 12.11 g of Tris base, 5.84 g NaCl, and 1.02 g of $MgCl_2 \cdot 6H_2O$ in 500 ml of distilled water. Adjust pH to 9.5, and bring volume to 1 liter with distilled water. Store at 4° C; the solution is stable for a year or more.

PROCEDURE

1. Run SDS-PAGE or IEF gel as previously described (see Chapter 4). Do not stain the gel.
2. Soak the gel for 30 min in Towbin's transfer buffer [20 mM Tris-Cl, 150 mM glycine, pH 8.0, 20% (v/v) methanol].
3. Wearing gloves, cut a sheet of nitrocellulose and four sheets of Whatman filter paper to the size of the gel. Soak the nitrocellulose in distilled water by floating it on the surface of the water for 5 min, and then submerge it in the water for an additional two min. Transfer the nitrocellulose to a container of transfer buffer, and soak it for 5 min. Soak the filter paper in transfer buffer as well.
4. Assemble the transfer sandwich as described in the manufacturer's user manual for the transfer tank. A generalized sandwich assembly for transfer is shown in figure 5.2.
5. Place the assembled sandwich into the transfer tank, making sure that the nitrocellulose is closest to the anode. Place the tank on a magnetic stir plate in a cold room. Fill the tank with transfer buffer, begin stirring, and attach the electrodes to the power supply.
6. Transfer the proteins from the gel to the nitrocellulose by electrophoresis at 100 V (constant voltage) for 1 h or at 30 V overnight.
7. (Optional) Disassemble the transfer sandwich and confirm transfer by staining the nitrocellulose for total protein with Ponceau S dye (available from Sigma):
   (a) In a plastic dish, weighing boat, or other container, cover the nitrocellulose with a solution of 0.5% (w/v) Ponceau S in 1% (v/v) glacial acetic acid. Incubate for 10 min with agitation.
   (b) Destain with distilled water for 2 min.

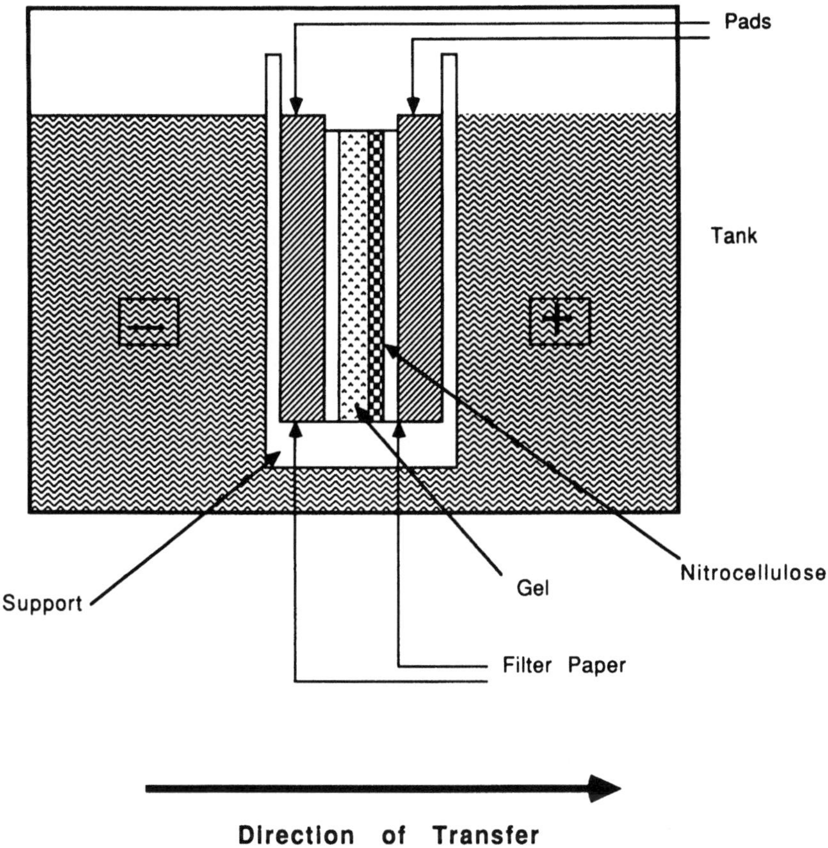

Figure 5.2 Schematic diagram of an electrophoretic transfer device used for transferring proteins from acrylamide gels to nitrocellulose sheets for Western blotting.

(c) Mark the location of the pink-stained protein bands with a small dot using a ball-point pen (Paper Mate ink works best for this).

(d) Completely destain with distilled water for 10–15 min.

8. Cover the nitrocellulose blot with blocking solution and agitate for 1 h.
9. Rinse the blot with TBS for 5 min. Repeat the wash twice more.
10. Cover the blot with about 8 ml of an appropriate dilution (see notes) of the primary antibody in TBS containing (0.5% (w/v) BSA. Agitate for at least 1 h (overnight incubations can be done).

11. Wash the blot with three changes of TBS for 5 min each.
12. Cover the blot with an appropriate dilution of the alkaline phosphatase-labeled secondary antibody (available from Bio-Rad). The final concentration of secondary antibody should be 1–5 μg/ml. Agitate for 1–2 h.
13. Wash the blot with four changes of TBS for 5 min each.
14. Prepare fresh developing solution by adding 66 μl of NBT stock to 10 ml of alkaline phosphatase buffer and mixing well. Add to this 33 μl of BCIP stock. Use this solution within 1 h.
15. Place the blot in a clean weighing boat or other container, and cover it with 10 ml of developing solution. Agitate until the bands appear well above the background.
16. Stop the reaction by removing the developing solution, and washing the blot with PBS/EDTA.
17. Photograph the blot to obtain a permanent record. The color fades within a few hours if exposed to room light.

A typical Western blot is shown in figure 5.3.

SOME NOTES ON WESTERN BLOTTING

The extent to which one should dilute the primary antibody must be determined empirically. Harlowe and Lane (1988) recommend that the final primary antibody concentration be 1–50 μg/ml. This usually represents a 1/100 to 1/2,000 fold dilution of polyclonal antibody-containing sera, etc.

Some workers have found that using non-fat dry milk for blocking can occasionally interfere with the binding of some antibodies to their antigen. An alternative, but more expensive, blocking solution can be prepared with 3% (w/v) BSA in TBS. If one wishes to use BSA for blocking, one must ensure that the BSA is free of IgG contaminants (such as Sigma product A7638). A third commonly used blocking solution consists of 3% gelatin in TBS (the gelatin is solubilized by heating to 50° C). This blocking solution is recommended by Bio-Rad for use with their Western blotting apparatus. Other blocking solution formulations can be found in the text by Harlow and Lane (1988).

A common problem encountered with both Western blotting and dot blots (*vide infra*) is high background staining of the nitrocellulose filters. Some workers have reported that background increases with age of the

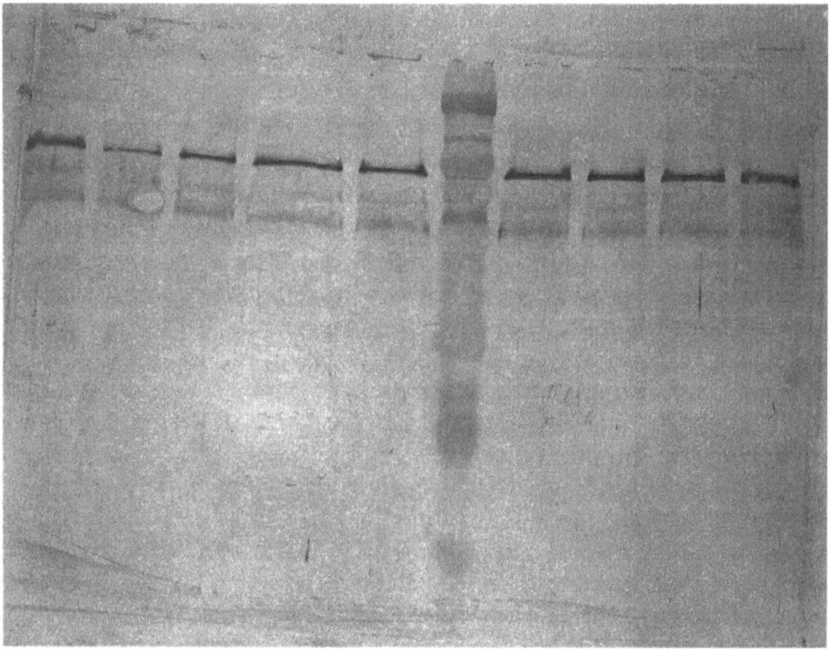

Figure 5.3 Example of a typical Western blot. Here four strains of *Rhodobacter sphaeroides* were grown aerobically and their cells collected and lysed. The cell lysates were subjected to SDS-PAGE and then transferred to nitrocellulose. This was then probed with an antibody raised against subunit II of the cytochrome *c* oxidase from *Paracoccus denitrificans*. In each of the four lanes (representing the four strains) one band with strong reactivity towards the antibody is observed. This band migrates at the expected molecular weight for the subunit II of cytochrome *c* oxidase from *Rhodobacter sphaeroides*. Data provided by Carol Wilson.

primary antibody, even when stored at −20° C. Another source of high backgrounds is insufficient blocking of the nitrocellulose prior to antibody exposure. If high backgrounds are encountered, one may wish to experiment with different blocking solutions and varying the length of time of blocking, until optimum conditions are found.

A relatively new detection method for Western blots and dot blots is enhanced chemiluminescence (ECL). This method used a secondary antibody that is labeled with horseradish peroxidase (HRP) to catalyze the oxidation of luminol, resulting in strong light emission. The emitted light can be used to expose photographic film, allowing a permanent record of the Western blot to be obtained. The method allows for detection of less than 1 picogram of antigen, and is safe and convenient to use. Amersham Corporation provides a kit for performing ECL-detected Western blots that includes the reagents, HRP-labeled secondary antibody, nitrocellulose membranes, and photographic film.

ANTIBODY CAPTURE ON MICROTITER PLATES

While Western blotting is extremely useful for confirming the identity of a protein band from a gel, there are times when one wishes to quantitatively screen large numbers of samples for the presence of a specific protein. Antibody capture on microtiter plates provides a relatively simple means of assessing the amount of antigen in various solution samples, such as fractions from a chromatographic purification step. In this method, samples that putatively contain the antigen (i.e., target protein) are adhered to the bottom of wells on a polystyrene microtiter plate. Unoccupied sites on the well surfaces are coated with a blocking solution, and then the primary antibody is added to the wells. A labeled secondary antibody is then used to detect adhesion of the primary antibody, after washing the wells to remove non-specifically bound antibody. Detection of antigen in this method is similar to that discussed for Western blotting (e.g., radioactivity, biotinylation, and enzyme-based chromogenic methods), except that here one wishes the detectable product to be soluble in aqueous solution. Using chromogenic detection methods, one can use this technique to determine quantitatively the amount of antigen in any given well, with the aid of a microtiter plate reader.

Here again, we shall describe in detail a method based on alkaline phosphatase detection, since this is a widely used and convenient method. Other detection methods are well described by Harlow and Lane (1988), Bjerrum and Heegard (1988), and in various product literature (e.g., Bio-Rad bulletin 1721).

PROCEDURE

1. Add 50 μl of each sample to be assayed to the bottom of a separate well on a polystyrene microtiter plate (available from Beckman, VWR, etc.). Place the plate on an orbital rocker and agitate at room temperature, in a humid environment, for at least 2 h.
2. Wash the wells by filling them with TBS. Remove the TBS by inverting the plate over a waste container and shaking down once to expel the liquid. Repeat the wash a second time.
3. Fill each well with BSA blocking solution (3% (w/v) BSA in TBS), and incubate with agitation for at least 2 h. (incubation can be done overnight).

4. Wash the plate as in step 2.
5. Add to each well 100 μl of primary antibody solution (at a concentration of ca. 1 μg/ml). Incubate with agitation for 2h.
6. Wash plates as in step 2.
7. Add to each well 100 μl of a 1:8,000 dilution of commercial IgG-alkaline phosphatase (making sure to use the correct species as described above). Incubate with agitation for 1–2 h.
8. Wash the plate four times as in step 2.
9. Wash the plate twice with 10 mM diethanolamine, 0.5 mM $MgCl_2$, pH 9.5, using the same method as in step 2.
10. To each well add 50 μl of substrate solution [1 mg/ml disodium p-nitrophenylphosphate in 10 mM diethanolamine, 0.5 mM $MgCl_2$, pH 9.5]. Cover the plate with aluminum foil to keep dark, and incubate with agitation for 30 min.
11. Stop color development by adding 50 μl of 0.1 M EDTA to each well.
12. Wells containing antigen will turn bright yellow. Read the absorbance of the wells at 405 nm.

The relative amount of antigen in each well can be assessed from the $A_{405\ nm}$ values obtained from the plate reader. To make this assay quantitative, one can include in the assay a set of wells containing a serial dilution of pure antigen (target protein) of known concentration. The concentration of antigen in any well is then determined from its value of $A_{405\ nm}$ by comparing it to the linear plot of $A_{405\ nm}$ vs. antigen concentration for the standard wells.

A number of variations on the antibody capture method have been reported. These include methods for screening samples for antibody content by use of a stock solution of antigen, and competitive binding assays to assess the relative avidity of antigens for a particular antibody or vice versa. For a more comprehensive description of these methods, the reader is referred to the text by Harlow and Lane (1988).

DOT BLOT ASSAYS

Dot blot assays provide a rapid and convenient means of screening large numbers of samples for the presence of antigen. The basis of the methods is that protein samples are bound to a nitrocellulose matrix. Free sites on the nitrocellulose are blocked with BSA, as described above for Western blots, and then a primary antibody, raised against one's target protein, is applied to the nitrocellulose. A labeled secondary antibody is then used to probe for binding of the primary antibody. This method is more convenient to perform on large numbers of samples than

Western blots, and can be used in a semi-quantitative way by observing the relative intensity of the dots that result from label development.

PROCEDURE

1. Wear gloves and use forceps for handling nitrocellulose throughout. Prepare a sheet of nitrocellulose (available from Schleicher and Schuell, Millipore, Bio-Rad, Whatman, etc.) by drawing a 4×4 mm grid of squares on the membrane with a pencil. Place the membrane in a plastic weighing boat, or other small container, and wash by immersion in distilled water for 5 min with agitation on an orbital rocker. Remove the membrane and dry it at room temperature.
2. For each sample to be tested, place a dot of 5 μl of sample (1–50 μg/ml in protein) on the center of one of the grid squares, and allow this to dry at room temperature. If the antigen samples are very dilute, multiple dotting of the same site on the nitrocellulose can be performed to build up the concentration of protein. If this is done, care should be taken to thoroughly dry the membrane between applications. Commercial nitrocellulose can bind up to ca. 100 μg protein per cm^2.
3. Place the membrane in a clean weighing boat, and wash it three times with TBS.
4. Block the membrane with 3% BSA blocking solution (3% (w/v) BSA in TBS with 0.02% azide) for at least 2 h (overnight blocking can be done).
5. Remove the wet membrane from the weighing boat, and place it on a sheet of parafilm. Dot 1 μL of primary antibody solution (ca. 50 μg/ml) onto the center of the grid squares (overlaying where the putative antigen solutions were applied). After a minute or so, overlay the nitrocellulose with blocking solution and incubate at room temperature for 30 min.
6. Wash the blot with three changes of TBS for 5 min. each.
7. Cover the blot with an appropriate dilution of the alkaline phosphatase-labeled secondary antibody (available from Bio-Rad). The final concentration of secondary antibody should be 1–5 μg/ml. Agitate for 1–2 h.
8. Wash the blot with four changes of TBS for 5 min each.
9. Prepare fresh developing solution by adding 66 μl of NBT stock to 10 ml of alkaline phosphatase buffer, and mixing well. Add to this 33 μl of BCIP stock. Use this solution within 1 h.
10. Place the blot in a clean weighing boat or other container, and cover it with 10 ml of developing solution. Agitate until the dots appear well above the background.

11. Stop the reaction by removing the developing solution, and washing the blot with PBS/EDTA.
12. Photograph the blot to obtain a permanent record. The color fades within a few hours if exposed to room light.

Figure 5.4 shows an example of the results of a dot blot assay. This particular example shows a high background, making it difficult to observe the dark dots that represent a positive result. Such high backgrounds usually result from insufficient blocking of the nitrocellulose

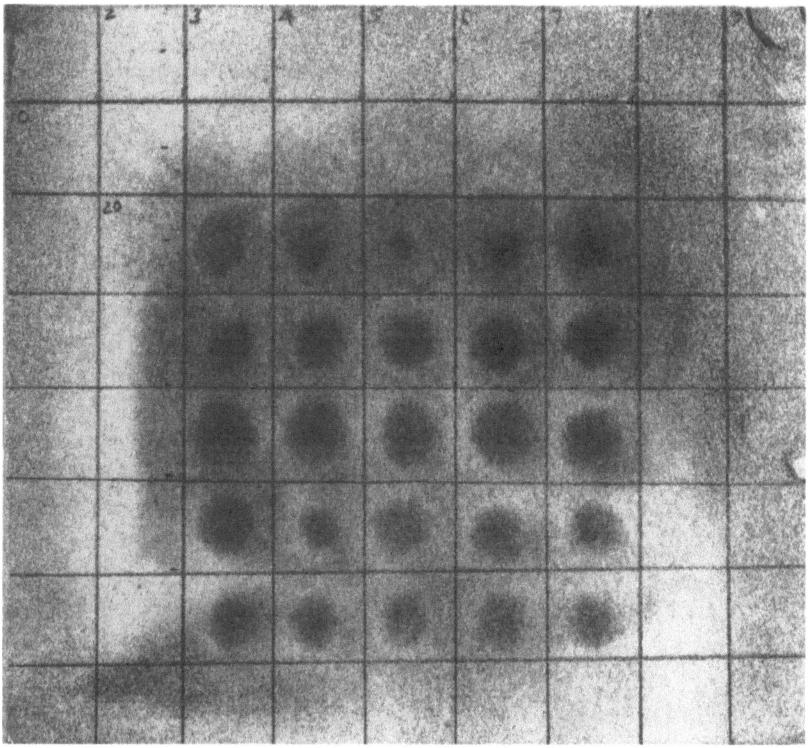

Figure 5.4 Example of a dot blot assay. Proteins from the periplasmic membranes of *Paracoccus denitrificans* cells were subjected to ion exchange chromatography, and some of the fractions were spotted onto nitrocellulose. The nitrocellulose was then probed with an antibody raised against subunit II of the cytochrome *c* oxidase from this organism. Note the high background in this figure. This is an indication that insufficient blocking of the nitrocellulose occurred after application of the samples. Better results could be obtained by increasing the incubation time in blocking solution or trying alternative blocking solutions.

after application of the antigen solutions. As with Western blotting, the solution to this problem is to incubate the nitrocellulose for longer times in the blocking solution, and/or experiment with alternative blocking solutions.

For antibody capture and dot blots no attempt to separate the target protein from other proteins in the sample is made. Thus, the utility of these methods depends on having an antibody with high selectivity for one's target protein (i.e., little cross-reactivity with other proteins). It is usually a good idea to test the selectivity of an antibody by Western blotting before using it for antibody capture and dot blot methods.

References

Andrews, A. T. (1986) *Electrophoresis: Theory, Techniques and Biochemical and Clinical Applications,* 2d ed., Oxford University Press, Oxford.

Axelsen, N. D. (1975) *Quantitative Immunoelectrophoresis: Methods and Applications,* Blackwell Scientific, Oxford.

Bjerrum, O. J., and Heegaard, N. H. H. (1988) *Handbook of Immunoblotting of Proteins,* CRC Press, Boca Raton, FL.

Bollag, D. M., and Edelstein, S. J. (1991) *Protein Methods,* Wiley, New York.

Harlow, E., and Lane, D. (1988) *Antibodies: A Laboratory Manual,* Cold Spring Harbor Laboratory, Cold Spring Harbor, New York.

Towbin, H.; Staehelin, T.; and Gordon, J. (1979) *Proc. Natl. Acad. Sci. USA,* **76,** 4350–4354.

6

Detection of Non-Protein Components

Within their natural biological environments, proteins often interact with other biological macromolecules, such as sugars, lipids, and nucleic acids. Sometimes these interactions are covalent, as in post-translational modifications of proteins, but more often involve non-covalent complex formation between the species. It is possible then, that non-protein components may occur in a protein sample, either as part of the biologically relevant active complex, or as contaminants that are not removed during purification. In this chapter we shall discuss methods for assessing the presence of the three most commonly occurring non-protein components in protein samples: nucleic acids, lipids from cell membranes, and covalently attached sugars in glycoproteins.

NUCLEIC ACIDS

Nucleic acids can sometimes occur in purified protein samples because of insufficient removal during isolation, specific interactions with the target protein, or leaching from nucleic acid-based affinity columns in the case of purification of nucleic acid-binding proteins. As we have seen in Chapter 3, the purine and pyrimidine bases of both DNA and RNA absorb light in the 260 nm region. When large amounts of nucleic acids (i.e., $>$ 250 ng/ml) are present in a sample, their absorbance properties will tend to distort the characteristic 280 nm band of proteins, by filling in the trough near 260 nm that is otherwise seen. Thus, a quick way of detecting the presence of large quantities of nucleic acids is to run an absorption spectrum of the protein sample between 250 and 350 nm, and compute the absorbance ratio at 260/280 nm (see Chapter 3). However, this method is rather insensitive and difficult to quantitate.

The most common methods for detection of nucleic acids are based on the abilities of certain aromatic molecules to interchelate between bases in the helical structures of oligonucleotides. Some of these molecules display strong fluorescence when complexed with the nucleic acids, thus leading to a sensitive and quantitative means of assessing the presence of these macromolecules in solution (Harris and Bashford, 1987). Ethidium bromide (EtBr) is the most widely used fluorescent interchelating agent for detection of nucleic acids. This reagent has been used for quantitative assays in solution, and for staining nucleic acids in agrose gels. Ethidium bromide fluoresces when interacting with either DNA or RNA, although the fluorescence enhancement is different for the two biopolymers. Another interchelating agent, 4′, 6-diamidino-2-phenylindole·2HCl (DAPI) is specific for DNA and provides even greater sensitivity than EtBr. Below we shall describe methods for the detection of DNA and RNA in protein samples, based on the fluorescence enhancement for both of these interchelators (Legros and Kepes, 1985).

Fluorometric Microassay for DNA

MATERIALS

1. Stock solution of 1.5 mg/ml DAPI (available from Molecular Probes, Eugene Oregon) in distilled water
2. Stock solution of calf thymus, *E. coli,* or other pure DNA sample (1 mg/ml) in the same buffer as the protein sample
3. Protein sample
4. Buffer.

PROCEDURE

1. In each of seven test tubes combine 200 μl of DAPI solution, 0, 100, 200, 300, 400, 500, and 600 μl of DNA stock, respectively, and enough buffer to bring the total volume to 3 ml. Mix well. This will give a set of solutions of DNA concentration ranging from 0 to 2.0 μg/ml and constant DAPI concentration of 90 ng/ml.
2. Read the fluorescence of these solutions at 452 nm with excitation at 346 nm (see Chapter 9 for information on obtaining fluorescence spectra).

3. Construct a standard curve by plotting the observed fluorescence intensity as a function of DNA concentration, similar to those constructed in Chapter 3 for protein quantitation.
4. Next prepare a set of samples of your protein solution in the same way as before representing a 1/100, 1/10, 1/2, and 0 dilution of your sample.
5. Record the fluorescence of these solutions and determine the DNA content of your samples using the standard curve. If the fluorescence of the more concentrated protein solutions are outside the linear range defined by the standard curve, disregard these samples.

NOTES ON THIS PROCEDURE

1. DNA concentrations as low as 1–5 ng can be detected in this way. Concentrations above 2 μg/ml tend to deviate from the linear standard curves.
2. Reagents such as SDS, EDTA, etc., can interfere with the assay. Make sure that all components of the buffer used for your protein are also present (in equal amounts) in your standard curve samples.
3. ***WARNING:*** *DNA-binding dyes are in general extremely carcinogenic. Use caution in handling them and always wear gloves and a lab coat when working with these reagents.*
4. Other dyes can be used in place of DAPI. For example, Hoechst 33258 is highly specific for DNA. See the Molecular Probes catalog (Eugene, Oregon) for other examples.

Pierce provides a DNA assay kit that uses a similar fluorometric assay as the one described here, but based on the DNA binding reagent bisbezimide (Hoechst 33258; Downs and Wilfinger, 1983). This kit (product number 23265 G) provides all of the reagents, including DNA standards, needed for performing about 100 assays. The manufacturer claims that the assay is sensitive enough to detect less than 3 ng of DNA.

The above assay is specific for DNA. There are times, however, when one wishes to assess the content of RNA in a sample as well. One method for detecting RNA is to use a non-specific nucleic acid-binding dye (one that binds both DNA and RNA) in conjunction with a DNA degrading enzyme.

Fluorometric Microassay for RNA

MATERIALS

1. Stock solution of 100 μg/ml ethidium bromide in buffer (PBS is a good buffer)

2. Stock of pure RNA (various species sources available from Sigma)
3. Stock solution of 1mg/ml DNase I (RNase free)
4. Stock solution of 1M $MgCl_2$
5. Buffer
6. Protein sample.

PROCEDURE

1. Prepare a set of calibration standards, as above, using the pure RNA. These should be 3.0 ml total volume, containing 0–5 μg RNA, 1 μg/ml DNase, 15 mM $MgCl_2$ and 100 μl (3.3 μg/ml) of stock ethidium bromide solution.
2. Measure the fluorescence of 580 nm for these solutions with excitation at 360 nm.
3. Construct a standard curve by plotting the observed fluorescence intensity as a function of RNA concentration.
4. Prepare a set of dilutions of your sample as per the DAPI assay. These should contain all of the reagents listed for the calibration standards (except pure RNA) and should have a total volume of 3.0 ml.
5. Measure the fluorescence of these samples and compute the RNA content, as described above for the DNA assay.

NOTES ON THIS PROCEDURE

1. EtBr fluoresces when bound to either DNA or RNA. The DNase I is used here to degrade any DNA which would otherwise severely interfere with the assay for RNA. Many grades of DNase are commercially available. It is *absolutely imperative* that one use an *RNase free* grade (available from Sigma), otherwise degradation of the RNA will occur and alter the results.
2. This assay can be adapted to simultaneously determine both RNA and DNA by preparing a duplicate set of samples in which the DNase I is replaced with RNase (DNase free grade).
3. The same warning applies here as for DAPI. *EtBr is extremely carcinogenic. Use caution!*

LIPIDS

Phospholipids are the main components of biological membranes. They are largely hydrophobic molecules with polar (hydrophilic) head groups. Because of their low water solubility, lipid contamination is usually not a problem for soluble proteins. For detergent-solubilized membrane proteins there is almost always some level of lipid contamination (in some cases removal of the trace lipids adversely affects enzyme activity). In these cases one needs to determine the quantity and type of lipid present in the sample. Lipids are separated from other components by extraction with non-polar organic solvents. A polar solvent (such as methanol) is usually included to help disrupt membrane-protein interactions, and acidic conditions are used in order to ensure complete extraction (Radin, 1969).

Separation of Lipids

MATERIALS

1. Chloroform/methanol (2:1 v/v) containing 0.25% concentrated HCl
2. Protein sample
3. *Glass* test tubes
4. *Glass* separatory funnel
5. 0.05 M $CaCl_2$ in distilled water
6. Table top centrifuge.

PROCEDURE

1. In a glass test tube dilute the protein sample with 20 volumes of chloroform/methanol and mix well.
2. Cover the tube and let it stand for 20 min.
3. Centrifuge in a table top centrifuge for 5 min and remove any precipitated protein.
4. Add 1/5 volume of $CaCl_2$ solution and mix well to form an emulsion. Pour this quickly into a separatory funnel and allow to stand until two well-defined phases are observed. The lower phase contains the extracted lipids in chloroform. Remove this phase and wash it with methanol/water (1:1, v/v) twice (each time removing the lower chloroform phase by use of a separatory funnel).
5. Remove the chloroform by evaporating using a stream of nitrogen in a fume hood.

The resulting lipid film can be analyzed by mass spectroscopy, TLC, infrared spectroscopy, Raman spectroscopy, or other appropriate methods depending on the quantity of sample. Since most workers will have access to TLC equipment, we shall detail this detection method here.

TLC detection (Skipski and Barclay, 1969)

The sample can be solubilized in chloroform (to a concentration of ~10 mg/ml) and spotted on a standard TLC plate (silica gel H, Merck). Separate spots of commercially available pure lipids should be run on the same plate (pure lipid samples can be obtained from Sigma and other suppliers). The R_f values can drift a bit so it is best to run standards and samples at the same time. Because different classes of lipids migrate very differently, depending on their hydrophobicity, etc., it is often best to run TLC plates using more than one solvent system. Table 6.1 lists appropriate solvent systems for different lipid types.

Visualizing lipid spots on TLC plates.

IODINE VAPOR

This is the most general and simple method for detection.

1. Remove plate from development tank and allow it to dry in a fume hood.

Table 6.1 Some solvent systems for separating membrane lipids.

Lipid Type	Solvent System
Neutral lipids	heptane/diethylether/glacial acetic acid 60/40/1
Neutral lipids	heptane/diethylether/glacial acetic acid 80/20/1
Phospholipids	chloroform/methanol/glacial acetic acid/water 25/15/4/2
Phospholipids	chloroform/methanol/glacial acetic acid/water 60/50/1/4
Phospholipids	n-propanol/propionic acid/chloroform/water 3/2/2/1
Acidic phospholipids	chloroform/methanol/glacial acetic acid/water 80/13/8/0.3
Polyphosphoinositides	n-propanol/4 M ammonia 2/1
Polyphosphoinositides	chloroform/methanol/4 M ammonia 9/7/2
Polyphosphoinositides	chloroform/methanol/28% ammonia/water 40/48/5/10
Gangliosides	chloroform/methanol/2.5 M ammonia 60/40/9
Gangliosides	propanol/water 7/3

2. After drying place the plate in a sealed container (another developing tank and lid works well) with a few crystals of iodine. Place in a warm spot to facilitate sublimation of the iodine.
3. Observe yellow-brown staining of lipid spots.

PHOSPHOLIPID STAIN

Phospholipids are by far the most plentiful lipid components of biological membranes. A very sensitive method for detecting phosphate-containing lipids is given below:

SOLUTION A

16 g ammonium molybdate in 120 ml H_2O.

SOLUTION B

40 ml concentrated HCl, 10 ml mercury plus 80 ml solution A. Mix vigorously for ~ 30 min and filter.

SOLUTION C

Add all of solution B and remainder of solution A to 200 ml concentrated sulfuric acid. Bring volume to 1 liter with distilled water and store in an *amber glass* bottle.

1. Remove the plate from developing tank and allow it to dry. Place dried plate in a cardboard box standing up in back of the fume hood and spray evenly with solution C.
2. Observe dark blue staining of phospholipid spots.

Note: Many lipids (in particular unsaturated lipids) are degraded by air oxidation. It is therefore best not to store lipids for long times. Also, solutions to be used for lipid isolation and detection can be deoxygenated by bubbling nitrogen through them.

PROTEIN GLYCOSYLATION

Many types of post-translational covalent modifications of proteins have been detected including covalent attachment of alkyl groups (methylation), lipids, sugars, and phosphate groups. One of the most commonly encountered covalent modifications of proteins from eukaryotic sources is the attachment of polymeric sugars in a process known as glycosylation to produce glycoproteins. Glycosylation plays important biological roles in protein transport, adhesion to cell membranes, protection from proteolytic degradation, and other processes *in vivo*. The presence of the carbohydrate moieties can also dramatically affect the physico-chemical properties of proteins *in vitro*, such as sample viscosity, sample adhesion to surfaces, and the apparent molecular weight. The carbohydrate portion of a glycoprotein can make up anywhere from 1 to over 80 percent of the total molecular weight of the molecule! Proteins can be glycosylated by covalent modification at the hydroxyl groups of serine or threonine residues (known as O-linked glycosylation), at the side chain of asparagine residues (known as N-linked glycosylation), or in the specific case of collagen at hydroxy-lysine residues. For O-linked glycosylation, no specific amino acid sequence is required to define a potential site of modification. For asparagine residues, N-linked glycosylation occurs for the specific sequences Asn-X-Ser or Asn-X-Thr, where X is any amino acid other than proline. The types of sugar constituents found in glycoproteins include galactose, mannose, glucose, frucose, xylose, N-acetylglucosamine, N-acetylgalactosamine, and sialic acids (Hughes, 1983). Methods for detecting the presence of oligosaccharides on proteins have been developed that involve the use of sophisticated gas chromatographic, mass spectral, and HPLC chromatographic techniques. These have been reviewed elsewhere (Hugli, 1989), but such methods are probably best left to experts. For the non-specialist, electrophoretic methods provide the most straightforward means of assessing whether or not a protein is glycosylated. Two methods are described below that allow one to probe for glycosylation of proteins after separation by gel electrophoresis. These are: carbohydrate staining (Leach et al., 1980), and lectin blotting (Olmstead, 1981). A third method, gel shift assays after enzymatic deglycosylation, will be briefly discussed.

Carbohydrate Staining

The procedure of McGuckin and McKenzie (1958), as modified by Leach et al. (1980) is commonly used to stain selectively glycosylated protein bands after gel electrophoresis. The proteins must be fixed in the gel prior to staining. Staining is accomplished by oxidation of the glycols to aldehydes with periodic acid, and then modification of the aldehydes with the chromogenic reagent pararosaniline.

MATERIALS

1. Apparatus for gel electrophoresis (see Chapter 4)
2. 5% (w/v) phosphotungstic acid (available from Sigma) in 2 N HCl

3. 7% (v/v) methanol, 14% (v/v) acetic acid in distilled water
4. 1% periodic acid in 7% trichloroacetic acid (available from Sigma)
5. 0.5% (w/v) metabisulfite in 0.1 NHCl
6. Schiff's reagent (available from Sigma, catalog number 395-2-016).

PROCEDURE

1. Run a standard SDS gel as described in Chapter 4.
2. Remove the gel and place in a large Pryex casserole dish, and cover with 5% phosphotungstic acid solutions. Agitate gently on an orbital rocker for 1.5 h.
3. Decant the phosphotungstic acid solution, and soak the gel in 500 ml of 7% methanol, 14% acetic acid solution with agitation for 1h. Decant this solution and replace with a fresh 500 ml of the same solution. Soak with agitation for another hour. This step removes excess SDS, which is critical for the analysis.
4. Decant the solution, and cover the gel with 1% periodic acid in 7% trichloroacetic acid. Cover the dish with aluminum foil to keep the gel dark, and agitate for 1 h. This step oxidizes the glycols to aldehydes.
5. Decant the solution and cover the gel with 300 ml of 0.5% metabisulfite in 0.1 N HCl. Agitate for 1 h. This step reduces excess periodic acid. As soon as the metabisulfite is added, one may notice an amber color develop in the gel as I_2 is formed. This color will disappear as the iodine is reduced to iodide.
6. Decant the solution and transfer the dish to a cold room or place on ice. Cover the gel with Schiff's reagent, and cover the dish completely with aluminum foil to exclude light (if possible turn off the room lights in the cold room as well). Allow the color to develop overnight at 4° C in the dark. The glycoprotein bands will be stained pink, and the color is stable for a few days. As little as 3 μg of glycoprotein can be detected by this method.

A modification of this method, in which the oxidized glycols are reacted with biotinylated hydrazine and then detected with streptavidin-alkaline phosphatase has recently been developed. A kit, based on this method is available from Oxford Glycosystems (Rosedale, New York) for detection of glycoproteins in gels and on dot plots. The manufacturer claims that one can detect ng quantities of glycoprotines with their kit.

An alternative carbohydrate staining method has been reported by Racusen (1979), based on treatment of the gel with thymol and sulfuric acid. The procedure follows.

MATERIALS

1. Apparatus for gel electrophoresis (see Chapter 4)
2. 25% (v/v) isopropanol, 10% (v/v) acetic acid in distilled water
3. 0.2 μg/ml thymol in 25% (v/v) isopropanol, 10 (v/v) acetic acid in distilled water
4. 80% (v/v) sulfuric acid, 20% (v/v ethanol).

PROCEDURE

1. Run a standard SDS gel as described in Chapter 4.
2. Remove the gel and place in a large Pyrex casserole dish, and cover with 100 ml of 25% isopropanol, 10% acetic acid in distilled water. Place on an orbital rocker and agitate for 2 h. Decant the solution and repeat this washing step.
3. Decant the solution and cover the gel with a solution of thymol, in the same isopropanol/acetic acid/water mixture as above. Agitate for 2 h.
4. Decant the solution and cover the gel with 100 ml of a solution of 80% (v/v) sulfuric acid, 20% (v/v) ethanol. Agitate until the glycoprotein bands appear pink/red against the yellow background of the gel (usually 2–4 h).
5. Photograph the gel immediately. The color fades after a few hours, and the background darkens.
6. If desired, the gel can be stained for total protein with Coomassie Blue. Soak the gel overnight in a solution of 10% acetic acid (v/v), 10% methanol (v/v) in distilled water. Once the gel is thus rehydrated, stain as described in Chapter 4.

This method is a little more sensitive than the pararosaniline staining method, but has the drawback that the stained bands fade rather quickly. Although these methods have been described for detection after electrophoresis, they can also be adapted to a filter binding assay (i.e., dot blot).

Lectin Blotting

Lectins are proteins that bind oligosaccharides tightly; perhaps the most well-known member of this class of proteins is concanavilin A from the Jack bean. Because of their tight and specific binding to oligosaccharides, lectins can be used to detect the presence of glycoproteins in much the same way that antibodies can be used to detect antigens. Lectin blotting, analogous to Western blotting, is commonly used to detect glycoproteins after separation by gel electrophoresis. Biotinylated lectins are commercially available, and form the basis for detection of the glycoproteins. After binding to the immobilized glycoprotein, the presence of the biotinylated lectin is detected with a streptavidin biotinylated enzyme conjugate. A commercial kit for this method is available from Vector Laboratories, Inc. (Burlingame, California). The following protocol, based on the Vector Laboratories kit, has been adapted from Casey et al. (1992).

MATERIALS

1. Apparatus for gel electrophoresis and transfer (see Chapters 4 and 5)
2. Nitrocellulose membranes
3. Blocking solution: 0.25% gelatin in 10% ethanolamine, 0.1 M Tris-Cl, pH 9.0
4. 0.2 μg/ml biotinylated tomato lectin (from Vector Laboratories) in 0.25% gelatin, 0.05% Noidet P-40 (from Sigma), 0.15 M NaCl, 5 mM EDTA, 50 mM Tris-Cl, pH 7.5
5. Streptavidin-biotinylated horseradish peroxidase stock (Vecta-Stain; from Vector Laboratories)
6. Diaminobenzidine Solution: dissolve 6 mg of 3,3′-diaminobenzidine in 10 ml of 50 mM Tris-Cl, pH 7.5. Filter the solution through Whatman filter paper to remove any small amount of precipitate. Prepare this solution fresh on the day of the blotting
7. 30% hydrogen peroxide solution.

PROCEDURE

1. Run the SDS-gel as described in Chapter 4, and electrophoretically transfer protein bands to nitrocellulose as described in Chapter 5.
2. Wash the membrane with TBS as described under Western blotting (Chapter 5), then block with the gelatin blocking solution for 2 h.
3. Wash the membrane with TBS as previously described.
4. Decant off the buffer and cover the membrane with 0.2 μg/ml biotinylated tomato lectin solution. Agitate overnight in a covered container.
5. Decant off the solution and wash the membrane in TBS as described above.

6. Cover the membrane with a 1:10,000 fold dilution of the Vecta-Stain stock of streptavidin-biotinylated horseradish peroxidase in blocking buffer. Agitate in a covered container for 18 h.
7. Wash the membrane in TBS as described above.
8. To 10 ml of diaminobenzidine solution add 10 μl of 30% H_2O_2, and mix well.
9. Decant the TBS and cover the membrane with the diaminobenzidine/peroxide solution prepared in step 8. Agitate the membrane until dark bands appear above the background (usually 1–5 min).
10. Stop the reaction by decanting the solution and washing the membrane with PBS.
11. Photograph the blot to obtain a permanent record.

Enzymatic Deglycosylation

Another potential method for detecting protein glycosylation is by comparing the electrophoretic mobility of a putative glycoprotein before and after treatment with a deglycosylating enzyme. Glycoproteins tend to display retarded migration in SDS-gels because of the increased molecular weight associated with the carbohydrate portion of the molecule, and also because of diminished SDS binding to these proteins; glycoproteins also tend to give broader, more diffuse bands than non-glycosylated proteins upon staining. Thus, it is often the case that these proteins will show a shift in migration on the gel after treatment with a deglycosylating enzyme. A number of deglycosylating enzymes are now commercially available that are specific for either N-linked or O-linked glycoproteins. For example, N-glycosidase F (available from Genzyme and Boehringer Mannheim) cleaves N-linked carbohydrates, while endo-N-acetylgalactosaminidase cleaves the O-glycan core of O-linked carbohydrates. Typically, one divides a protein sample into two aliquots. One of these is treated with a catalytic amount of the deglycosylating enzyme, while the other is left untreated. Both samples are then run on a denaturing SDS-gel, and their relative mobilities are compared. A difficulty with this method is that the degree to which band mobility is perturbed depends on the extent to which the sample is glycosylated. Thus, a failure to observe a shift in electrophoretic mobility does not necessarily rule out protein glycosylation. A more detailed description of this method, and protocols for the use of deglycosylating enzymes, can be found in the review by Gerard (1990).

OTHER POST-TRANSLATIONAL MODIFICATIONS

While protein glycosylation is the most widely encountered post-translational modification for proteins from eukaryotic sources, other forms of covalent modification are encountered. As described above, these modifications are quite varied in the chemistry involved (phosphorylation vs. alkylation, vs. lipidation, etc.) and hence vary in the analytical methods used for their detection. In general, each modification type requires a specific and rather sophisticated method of analysis that is outside the scope of this book, and outside the scope of the general protein laboratory. An excellent review of post-translational modifications of proteins is provided by Wold (1981), and descriptions of some methods germane in detection of these modifications are described in the text by Hugli (1989).

References

Casey, J. R.; Pirraglia, C. A.; and Reithmeier, R. A. F. (1992) *J. Biol. Chem.*, **267,** 11940–11948.

Downs, T. R. and Wilfinger, W. W. (1983) *Analyt. Biochem.*, **131,** 538–547.

Gerard, C. (1990) *Meth. Enzymol.*, **182,** 529–539.

Harris, D. A., and Bashford, C. L. (1987) *Spectrophotometry and Spectrofluorimetry: A Practical Approach*, IRL Press, Oxford.

Hughes, R. C. (1983) *Glycoproteins*, Chapman and Hall, New York.

Hugli, T. E. (1989) *Techniques in Protein Chemistry*, Academic Press, San Diego.

Leach, B. S.; Colawn, J. F., Jr.; and Fish, W. W. (1980) *Biochemistry*, **19,** 5734–5741.

Legros, M., and Kepes, A. (1985) *Analyt. Biochem.*, **147,** 497–502.

McGuckin, W. F., and McKenzie, B. F. (1958) *Clin. Chem.*, **4,** 476.

Olmsted, J. B. (1981) *J. Biol. Chem.*, **256,** 11955–11957.

Racusen, D. (1979) *Analyt. Biochem.*, **99,** 474–476.

Radin, N. S. (1969) *Meth. Enzymol.*, **14,** 245–254.

Skipski, V. P., and Barclay, M. (1969) *Meth. Enzymol.*, **14,** 530–598.

Wold, F. (1981) *Ann. Rev. Biochem.*, **50,** 783–814.

7

Peptide Mapping and Amino Acid Analysis

As we have seen in Chapter 1, what uniquely defines a specific protein is its amino acid sequence, or primary structure. The methodology for determining the sequential arrangement of amino acids in a protein or peptide has been in place for some time. The first protein for which a full amino acid sequence was determined was the peptide hormone insulin by Frederick Sanger in 1953. At the time, this work was considered a major breakthrough, since it proved that proteins have uniquely defined structures. Sanger received the Nobel Prize for this work in 1958. While much of the chemistry involved in amino acid sequencing is now automated, the determination of protein sequences remains a tedious and technically demanding effort that is best left to laboratories that specialize in these methods. Nevertheless, the generalist can gain insight into the structure of target proteins by a number of methods that provide indirect information of amino acid composition and arrangement. Peptide mapping and amino acid analysis are often combined to provide this type of information. In this chapter we shall describe the methods commonly employed for peptide mapping and amino acid analysis of proteins. As we shall see, these methods are readily accessible to most protein scientists. At the end of the chapter, we shall briefly discuss the basis for protein sequence analysis. Again, the actual determination of protein sequences is largely the realm of experts, but it is worthwhile reviewing the chemistry involved in sequence analysis, since this is such an important part of modern protein science.

PEPTIDE MAPPING

In peptide mapping one seeks to define the specific pattern of peptides that arises from treatment of a protein with a specific proteolytic enzyme or chemical (Mihalyi, 1978; Darbre, 1986; Beynon and Bond, 1989). As described in Chapter 2, proteases (or endopeptidases) are enzymes that cleave peptide bonds, thus breaking up the linear chain of a polypeptide or protein into smaller peptide fragments. Some proteases are very specific in terms of the amino acid sequences that they recognize as cleavage sites. Table 7.1 provides a list of some of the more commonly used proteases for peptide mapping, their recognition sequences, and other useful information about these enzymes. The specificity exhibited by these enzymes can be used by the protein scientist to great advantage. If two protein samples are thought to represent the same purified protein, then treatment of these samples with a specific protease should result in identical patterns of peptide fragments. The application of more than one protease, with differing cleavage sites, can be applied to different aliquots of a sample to further enhance the power of this method. Furthermore, if the amino acid sequence of the target protein is known, one can predict the number of proteolytic fragments, and their molecular weights, that should result from treatment of that protein with a specific protease. If the pattern of peptides obtained from a sample matches that predicted on the basis of the target protein's amino acid sequence, this is good, albeit indirect, evidence that the sample is the correct protein.

Sample Preparation

In order for the cleavage reagent to attack all possible target sites on the protein, one must ensure that the target sites are exposed to the agent. This generally means that the protein should be denatured (i.e., unfolded) and free of disulfide bonds. To ensure that disulfide bonds do not spontaneously reform, cysteine residues are covalently modified so that they are no longer reactive. The following procedure is useful for most proteins.

1. Dilute the protein to 10–20 mg/ml into a buffer containing 8 M urea, 0.5 M NH_4HCO_3, pH 8.0. Place the sample in a polypropylene tube and flush with nitrogen. Cap the tube and incubate for 30 min at 37° C (this step denatures the protein).

Table 7.1 Some examples of proteolytic enzymes that are useful for peptide mapping.

Enzyme	Optimum pH	Major cleavages	Minor cleavages	Exceptions
Trypsin	7–9	Lys-X, Arg-X		Lys-Pro, Arg-Pro
Thrombin	8	Arg-X (X=Gly, Ala, Val, Asp, Cys, Arg, His)		
S. aureaus V8 protease	4, 7.8	Glu-X	Asp-X	Glu-Pro, Glu-Glu
Clostripain	7.7	Arg-X		
Mouse sub-maxillary protease	7.5–8	Arg-X		
Post-proline cleaving enzyme	7.5–8	Pro-X		Pro-Pro
Chymotrypsin	7–9	H-X (H=Tyr, Phe, Trp, Leu)		H-Pro
Thermolysin	7–8	X-H (H=Val, Leu, Ile, Phe, Tyr, Trp)		X-H-Pro
Asparaginylendopeptidase	5	Asn-X		Asn-N terminus Asn-X-N terminus glycosylated Asn
Pepsin	2	X-H-Y,X-Glu-Y (H=aromatic or large aliphatic)		

2. Add dithiothreitol (DTT; available from Sigma) to a concentration of 5 mM, flush the tube again with nitrogen. Cap and incubate for 4 h at 37° C or 30 min at 50° C (this step reduces disulfide bonds).
3. Cool the tube on ice and add a slight molar excess over cysteine residues (i.e., over SH groups) of either iodoacetic acid or iodoacetamide (both available from Sigma). A useful rule of thumb, if one is not sure of the cysteine content, is to add 5 μl of 100 mM iodoacetamide for every 20 nanomoles of protein present. Adjust pH back to 8.0 if necessary, and incubate *in the dark* at 4° C for 1 h or at room temperature for 15 min (this step covalently modifies cysteine residues to form S-carboxymethyl cysteine).
4. After incubation, add excess β-mercaptoethanol to react any excess alkylating reagent and run the protein down a desalting column to remove denaturants (Sephadex G-10, Biogel P-10, etc., in 100 mM NH_4HCO_3).
5. At this point the protein is S-carboxymethylated, but may have partially renatured. Before proceeding with cleavage one must ensure complete unfolding. Place the protein (in a plastic container) in a boiling water bath and incubate for 5–10 min. Cool to room temperature.

At this point the protein should be completely unfolded. One potential problem at this stage is protein precipitation. If the protein precipitates, one can try resolubilizing in low concentrations of urea, guanidine-HCl, or SDS, depending on the cleavage agent to be used (some proteases are remarkably stable to treatment with these denaturants). Sonication of the sample for 5 min at 4° C in a bath sonicator is also sometimes helpful in solubilizing the sample. Similar methods of sample preparation using guanidine hydrochloride and 2-mercaptoethanol in place of urea and DTT, respectively, have also been reported (Flannery et al., 1989).

Once prepared the sample is usually treated by addition of protease to 1–2% (w/w) of the target protein mass (see Beynon and Bond (1989) for specific recommendations for different proteases). Proteolysis is generally carried out at 37° C for 2–4 h, but longer incubations can be performed. If overnight incubations are done, steps should be taken to ensure that the solution is free of bacterial contaminants. Long incubation times can also favor non-specific proteolysis due to trace amounts of other contaminating proteases; thus care should be taken to use only highly purified protease for these experiments. A staggering variety of proteases are now commercially available. Sigma, Calbiochem, Worthington, and Boehringer Mannheim each supply a large and diverse group of proteolytic enzymes, too many to be dealt with in detail here. Instead, we shall detail the conditions for use of a number of commonly used proteases that are useful for peptide mapping. Conditions for use of other proteases can be obtained from the individual suppliers.

Trypsin

Trypsin cleaves proteins on the C-terminal side of lysine and arginine residues, between Lys-X and Arg-X bonds, where X is any residue other than proline. When X is an acidic residue, the rate of proteolysis may be reduced, but cleavage still occurs. If X is another Lys or Arg residue, or the N-terminal residue, the rate of proteolysis is significantly reduced. Commercially available trypsin contains varying amounts of chymotrypsin as a contaminant. This later protease is inhibited by L-(1-tosylamido-2-phenyl)ethylchloromethyl ketone (TPCK). TPCK-treated trypsin is available commercially, and should always be used to ensure the specificity of cleavage. Trypsin performs best between pH 7 and 9, and is reversibly denatured below pH 4. Because trypsin is autodigestable (i.e., it will proteolyze itself), solutions of trypsin should be prepared immediately before use. Alternatively, stock solutions of trypsin (10 mg/ml) can be prepared in 10 mM HCl, and stored at −20° C for several weeks.

Digestion with trypsin is done by addition of the protease to 1–2% (w/w) of the mass of the target protein in 100 mM ammonium bicarbonate, with incubation at 37° C for 2–4 h. The reaction is stopped by freezing the sample, or by addition of a serine protease inhibitor, such as PMSF (Sigma) or soybean trypsin inhibitor (Sigma) at 4mg/mg of protease.

Chymotrypsin

Chymotrypsin cleaves proteins on the C-terminal side of aromatic and large hydrophobic residues, between H-X bonds, where H is Tyr, Phe, Trp, or Leu, and X is any amino acid other than proline. The conditions and procedures for digestion with chymotrypsin are the same as described above for trypsin. Chymotrypsin is also a serine protease, and can thus be inhibited with PMSF. The reaction can also be stopped by freezing the sample.

S. aureaus V8 Protease

This protease cleaves proteins on the C-terminal side of glutamate residues, between Glu-X bonds as long as X is not proline or another glutamate residue. The protease is stable over a wide pH range (3.5–9.5), but its specificity of cleavage is buffer dependent. For cleavage only after Glu, one should use 50 mM ammonium bicarbonate (pH 7.8) as the buffer. If instead, one uses 50 mM sodium phosphate (pH 7.8), cleavage will occur after both Glu and Asp residues.

Digestion is accomplished by addition of the protease to 1–2% (w/w) of the target protein mass, in one of the buffers described above (depending on the specificity one wishes to achieve), and incubation at 37° C for 4–18 h. Digestion is usually stopped by freezing the sample.

Thrombin

Thrombin is a protease of the serum fibrinolytic cascade that specifically cleaves proteins on the C-terminal side of arginine residues. Bonds of the type Arg-X are cleaved when X is Gly, Ala, Val, Asp, Cys, Arg, or His. Because of its specificity, thrombin is particularly useful for obtaining small numbers of large molecular weight fragments from a protein, and is thus commonly used in peptide mapping.

Digestion with thrombin is carried out in much the same way as described for trypsin. The optimal pH for thrombin activity is 8.0. Incubation is done at 37° C for 4–8 h. The reaction can be stopped by freezing the sample, or by inhibition of the enzyme with PMSF.

Clostripain

This protease also selectively cleaves Arg-X type bonds, but in some cases also cleaves Lys-X bonds. The enzyme requires a sulfydryl reagent for activity, and most workers use dithiothreitol for this purpose. A typical buffer system for clostripain proteolysis is: 100 mM ammonium bicarbonate, 1 mM DTT, pH 7.8. Digestion is performed at 37° C for 4 h with anywhere from 0.5 to 2% (w/w) clostripain to target protein ratios.

Asparaginyl-endopeptidase

This protease has only recently been isolated from the jack bean, and is now commercially available from TaKaRa Biochemical Inc. (Berkeley,

California; product number 7319). This enzyme cleaves specifically on the C-terminal side of asparagine residues, including Asn-Pro bonds. According to the supplier, the only exceptions are that Asn-X bonds are not cleaved when Asn is the N-terminal amino acid, or next to the N-terminal amino acid, or when the Asn residue is glycosylated. The enzyme is sold in activity units, rather than by mass, and comes in vials of 0.2 mU in 20 mM sodium acetate, 50% glycerol, 0.005% Brig-35 (a detergent), 1 mM DTT, 1 mM EDTA, pH 5.0. The enzyme is stable in this solution for at least a year, if stored at −20° C. Because this protease has only recently been isolated, there is not a large body of literature pertaining to it. The supplier recommends the following protocol for its use in digesting proteins.

Dissolve 10 nmol of the target protein in 100 μl of 20 mM sodium acetate, 10 mM DTT, 1 mM EDTA, pH 5.0. Add to this 0.1–0.2 mU of the enzyme, and incubate at 37° C for 15 h. Stop the reaction by adding acetic acid to a final concentration of 10% (v/v).

Because of the high specificity of this protease, it is likely to become an important tool for peptide mapping.

Cyanogen Bromide

Cyanogen bromide (CNBr) is a chemical, rather than enzymatic protein cleavage reagent that is widely used in peptide mapping. CNBr cleaves proteins on the C-terminal side of methionine residues, converting them to homoserine in the process. Unlike the catalytic reactions of enzymatic proteases, cleavage by CNBr requires large excesses of the reagent to achieve complete digestion, typically a 1,000-fold excess of CNBr over Met residues is used. CNBr is commercially available as a dry solid, or as a concentrated solution in acetonitrile (Aldrich). The reagent is volatile, and this fact must be kept in mind when using it. Whenever one uses CNBr, the work should be done in a properly functioning chemical hood. The following procedure is recommended for digestion of proteins with CNBr.

1. In a hood add some solid CNBr to a pre-weighted screw top vial, and seal the vial tightly. Re-weigh the vial, and determine, by difference, the mass of solid CNBr in the vial. Open the vial and quickly add the appropriate volume of 70% formic acid to bring the concentration of CNBr to 1 to 10 mg/ml. Reseal the vial tightly. This step should be performed immediately prior to digestion.

2. Dissolve a known mass of the target protein in 70% formic acid, and place it in a sealable container (for example, a plastic screw top vial).
3. Add a volume of the CNBr solution (from step 1) to the protein to bring the final CNBr concentration to two times the mass of protein. Reseal the container tightly.
4. Incubate the mixture at room temperature, in the dark, for 18–24 h.
5. Stop the reaction by adding 10 volumes of distilled water.
6. Remove the solvent and CNBr by lyophilizing the protein solution (see Chapter 2), or by blowing a stream of nitrogen gas over the top of the solution, within a hood, until dryness.
7. Dissolve the lyophilized protein in an appropriate buffer for later detection by SDS-PAGE or HPLC (see Chapter 4).

Detection of Peptide Fragments

Once digestion by a protease has been accomplished, one next needs a way of determining the pattern of proteolytic fragments that have been generated. The two most common methods used for this purpose are SDS-PAGE and reverse phase HPLC.

For SDS-PAGE separation one must consider the range of sizes that the proteolytic fragments are likely to display. Since these can be relatively low molecular weights, higher percentage acrylamide resolving gels are usually used (15 to 25% gels). Typically, one digests aliquots of the same sample with different proteases, and runs the resulting peptides in separate lanes of the same gel. One lane should also be run with a sample of the target protein that has not been subjected to digestion by any proteases, as a control. The procedure for electrophoresis, and staining of bands is the same as that described in Chapter 4. After electrophoresis, the pattern observed in the different lanes (corresponding to different protease treatments) can be compared to that of a standard target protein sample. A useful trick, to aid in making such comparisons, is to place the stained gel on a photocopying machine and make a photocopy onto clear acetate (overhead projection sheets). The acetate sheets with the photocopies of the sample digest and the standard protein digest can then be overlayed on a lightbox for easy comparison. Note that the quality of the photocopies, and hence the utility of this method, depends on the degree of staining achieved for the protein bands.

For small peptide fragments (5,000–1,000 Daltons), Swank and Munkres (1971) have suggested the use of a discontinuous urea-gel system using a number of strong, highly mobile electrolytes, as an alternative to the more common gel systems employed for higher molecular weight proteins. Andrews (1986) provides the following recipe for running such gels.

UPPER RESERVOIR BUFFER (CATHODE)

8.96 g (0.074 M) Tris Base

1.00 g (0.1% w/v) SDS

HCl to pH 7.8

Bring to 1 liter with distilled water

LOWER RESERVOIR BUFFER (ANODE)

24.22 g (0.2 M) Tris Base

0.40 g (0.04% w/v) SDS

H_2SO_4 to pH 7.8

Bring to 1 liter with distilled water

SEPARATING GEL BUFFER

2.42 g (0.2 M) Tris Base

0.04 g (0.04% w/v) SDS

48.05 g (8 M) urea

H_2SO_4 to pH 7.8

Bring to 100 ml with distilled water.

STACKING GEL BUFFER

2.42 g (0.2 M) Tris Base

0.04 g (0.04% w/v) SDS

H_2SO_4 to pH 7.8

Bring to 100 ml with distilled water.

SAMPLE BUFFER

0.17 g (0.139 M) Tris Base

0.1 (1% w/v) SDS

2.0 g (20% w/v) sucrose

70 μl (0.1 M) 2-mercaptoethanol
Acetic acid to pH 7.8
Bring to 10 ml with distilled water.
Store at −20° C in 100 μl aliquots.

PREPARING SEPARATING GEL

Dissolve 3.80 g of acrylamide and 0.20 g of bisacrylamide in 46 ml of separating gel buffer. Mix well, filter, and degas as described in Chapter 4. Add 25 mg of ammonium persulfate and swirl gently to mix. Add 25 μl of TEMED, and pour gel as described in Chapter 4.

PREPARING STACKING GEL

Dissolve 1.25 g of acrylamide and 0.31 g bisacrylamide in 48 ml of stacking gel buffer. Mix well, filter, and degas as described in Chapter 4. Add 25 mg of ammonium persulfate and swirl gently to mix. Add 25 μl of TEMED, and pour gel as described in Chapter 4.

PREPARING SAMPLES

Dissolve samples in sample buffer and heat for 1–5 min in a boiling water bath. Cool to room temperature, and load in gel wells as described in Chapter 4.

It is sometimes useful to recover individual peptide fragments from a gel for subsequent amino acid analysis or sequencing. This can be accomplished as follows. Place the gel on a clean glass plate, or other hard surface. With a clean razor blade or scalpel, cut out the band of interest. Again using the razor blade, chop the cut out band into small segments, and place these into a plastic test tube or other container. Cover the gel segments with 5 ml of 10 mM Tris-Cl, 0.01% SDS, pH 6.8. Cap the tube and incubate at 30° C, with agitation, overnight. Centrifuge the sample to pellet down the solid gel material. The peptide fragment will be solubilized in the supernatant (Andrews, 1986).

A common alternative to separation by SDS-PAGE is the use of reverse phase HPLC (see Chapter 4 for a discussion of protein HPLC). Typically a C8 or C18 column is used for peptide separation (Beynon and Bond, 1989; Darbre, 1986). Since proteolysis is usually done in votatile buffers, the solvent and buffer can be removed by lyophilization, leaving behind the peptide fragments, any unproteolyzed target protein, and the protease itself. These are then solubilized in 0.1% TFA in distilled water, or in aqueous 6 M guanidine hydrochloride, 0.1% TFA. The column is equilibrated with 0.1% TFA in distilled water, and after loading the sample, peptides are eluted with a linear gradient from 0 to 70% acetonitrile (as described in Chapter 4). Control samples containing only the unproteolyzed target protein, and another containing only the proteolytic enzyme should also be run so that the peaks from these components can be identified in the chromatograph resulting from the proteolysis sample. Peak detection is usually done by absorbance in the deep ultraviolet (205, 214, or 220 nm), so that peptides that do not contain tyrosine or tryptophan residues can be easily observed.

Determination of Disulfide Pairing by Peptide Mapping

An important aspect of protein tertiary structure is the arrangement of intramolecular disulfide bonds that are formed between specific pairs of cysteine residues. Which residues pair to form disulfide bonds in the native protein structure is a key issue in the analysis of proteins, since incorrect pairing can lead to loss of biological activity. The determination of disulfide bond arrangement is by no means a trivial task, but a common strategy for assessing this is to compare the pattern of peptide fragments obtained after proteolysis under reduced and non-reduced conditions.

Let us consider the hypothetical case of the protein, illustrated in figure 7.1, that contains a single disulfide bond and three cleavage sites for some protease. If this protein were reduced and its cysteine residues alkylated as described above, after proteolysis we would expect to observe four peptide fragments (corresponding to segments 1–4 in figure 7.1). If these were separated by reverse phase HPLC, we might observe the chromatograph shown in figure 7.2a. If, however, the protein were not reduced and alkylated prior to proteolysis, there would remain a covalent link (disulfide bond) between fragments 2 and 3. Instead of four peptide fragments we would now only observe three fragments, fragments 1 and 4 as before and a new fragment, 5, consisting of the disulfide linked fragments 2 and 3. The HPLC profile under these conditions might look like that shown in figure 7.2b. By comparison of the two chromatographs shown in figure 7.2 we could predict that the disulfide bond occurs between cysteine residues on fragments 2 and 3. If we have chosen a protease that results in fragments with no more than one cysteine per fragment, the information from the chromatographic comparison would suffice to uniquely identify the cysteine residues involved in disulfide bond formation. Of course, this example presents the

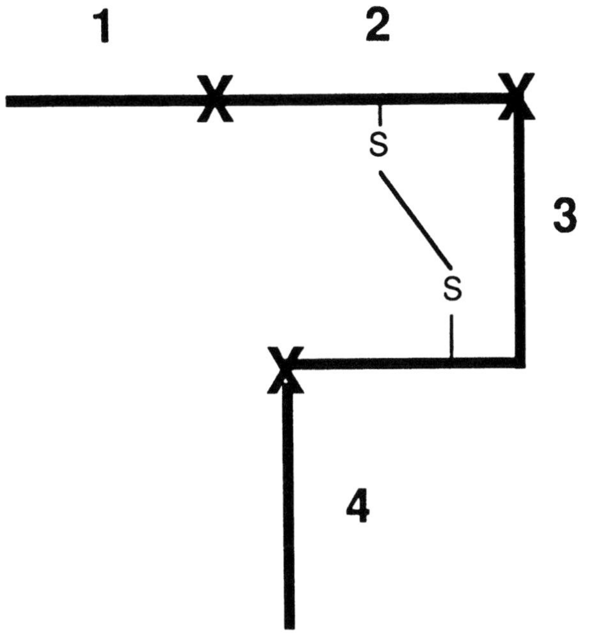

Figure 7.1 Diagram of a protein with three proteolytic cleavage sites (X) for some hypothetical protease. The four peptide fragments that would result from the action of this protease on the protein are labeled 1 through 4 here. Note that fragments 2 and 3 are covalently linked to each other through a disulfide bond.

simplest of cases. In real proteins, the analysis can be complicated by the presence of multiple disulfide bonds, and the inability of any protease to provide fragments with only single cysteine content. Nevertheless, this basic strategy is what is most commonly used to determine the pairing of cystine residues in intramolecular disulfide bonds.

An alternative method for detecting disulfides is *diagonal gel electrophoresis* (Creighton, 1974; Brown and Hartley, 1966). In this method one cleaves the protein into peptides with an appropriate protease, and then separates them by electrophoresis under non-reducing conditions (see Chapter 4). After electrophoresis the gel is soaked in performic acid, which cleaves the disulfides and converts Cys to $CysO_3H$ (very acidic). The gel is then electrophoresced at 90° to the original direction of migration. Those peptides that do not participate in disulfide bonding will be unaffected by the performic acid treatment, and will therefore migrate at the same rate as before treatment; these fragments will show up as a diagonal pattern on the gel. However, those peptides that have under-

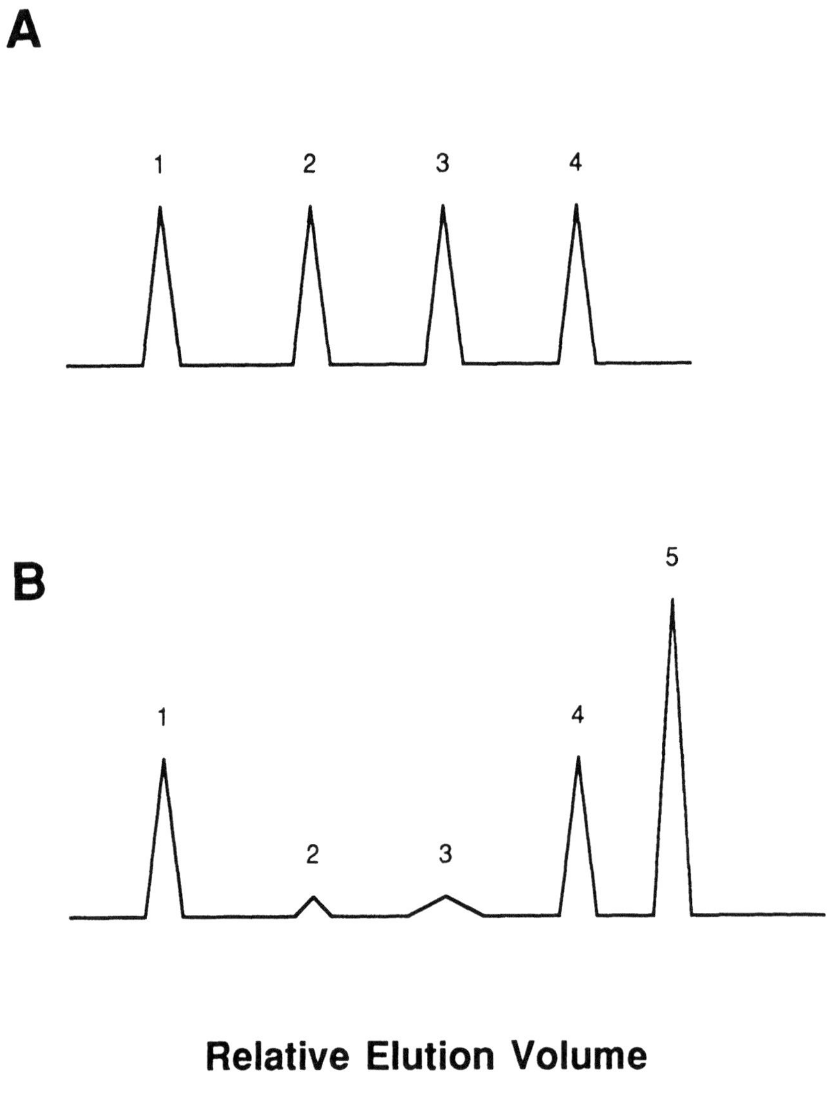

Figure 7.2 Cartoon of the expected elution profiles for the peptide fragments of the protein from figure 7.1 from a reverse phase HPLC column. In panel A the protein was reduced and alkylated prior to proteolysis, while in panel B the protein was not treated to reduce disulfide bonds.

gone conversion will migrate differently in the second dimension and thus show up as *off diagonal spots* (see figure 7.3). The off diagonal spots could then be cut out (*vide supra*) and subjected to amino acid analysis or sequencing to determine the pattern of disulfide bonds in the native protein.

A Simple Example of Peptide Mapping

As a simple illustration of the use of peptide mapping let us suppose that we have attempted to purify the mitochondrial protein cytochrome *c* from horse heart. We know the amino acid sequence of our target protein (Dayhoff, 1969), and from it we expect the protein to have a molecular weight of 12,300. Let us further imagine that we have gone through a purification scheme and obtained a sample that, when subjected to SDS-PAGE, gives a single band at approximately the correct molecular weight. Is this protein cytochrome *c*? The amino acid sequence for horse cytochrome *c* is shown in figure 7.4. After surveying this sequence we might decide that cleavage with cyanogen bromide (CNBr) and asparaginyl-endopeptidase (ASNase) would be useful, since both of these reagents should result in a relatively small number of peptide fragments. As shown in figure 7.4, CNBr treatment of this protein is expected to yield three peptides: CNBrl, MW = 7,760; CNBr2, MW = 1,810; and CNBr3, MW = 2,780. Likewise, ASNase treatment of cytochrome *c* should result in five peptides: A1, MW = 3,400; A2, MW = 2,310; A3, MW = 220; A4, MW = 1,760; and A5, MW = 3,630. Thus, we could take a sample of our protein, denature, reduce, and acetylate it, and then divide this sample into three equal aliquots. One aliquot would be left untreated as a control (uncut), while the other two would be treated with ASNase and CNBr, respectively. After proteolysis, all three samples could be run on the same SDS gel to determine the number and molecular weights of peptide fragments. If our sample were cytochrome *c*, we would expect to see an electrophoretic map similar to that shown in figure 7.5. More definitive proof of the identity of our sample protein could then be gained by amino acid analysis of selected peptide bands from our gel. If the amino acid composition of the peptide bands matched that expected for the proteolytic fragments of cytochrome *c*, this would be fairly good evidence that we have indeed isolated the correct protein.

AMINO ACID ANALYSIS

For routine screening of samples, it is often sufficient to observe the correct pattern of proteolytic fragments from peptide mapping, for identi-

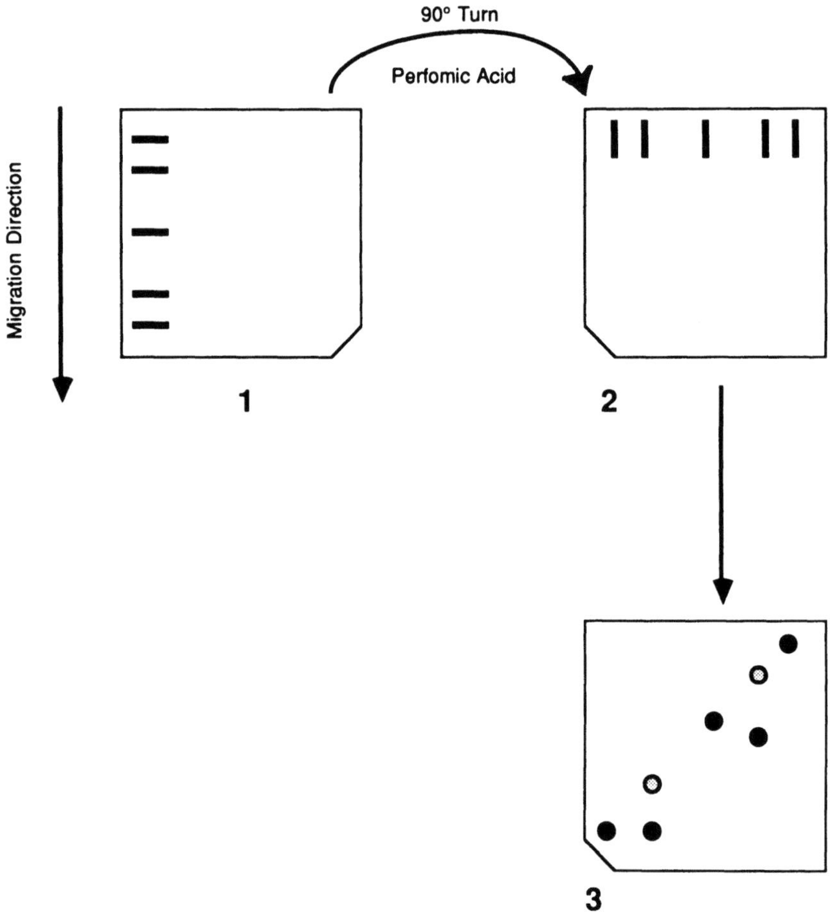

Figure 7.3 Steps in the performance of diagonal gel eletrophoresis for the determination of disulfide bonds between peptide fragments. After proteolysis under non-reducing conditions the peptide fragments are separated by SDS-PAGE in (**1**). The gel is then treated with performic acid as described in the text. After treatment the gel is loaded into an electrophoresis apparatus at 90° to the original direction of migration and electrophoresis is performed a second time (**2**). After staining, the resultant gel (**3**) shows a diagonal pattern of protein bands for peptides that were not affected by the performic acid treatment. Those peptides that were involved in disulfide bonding before treatment appear as off diagonal spots on the gel.

```
GLY-ASP-VAL-GLU-LYS-GLY-LYS-LYS-ILE-PHE-VAL-GLN-LYS-

                         A1
CYS-ALA-GLN-CYS-HIS-THR-VAL-GLU-LYS-GLY-GLY-LYS-HIS-

                       |
LYS-THR-GLY-PRO-ASN-LEU-HIS-GLY-LEU-PHE-GLY-ARG-LYS-

                         A2                        |
THR-GLY-GLN-ALA-PRO-GLY-PHE-THR-TYR-THR-ASP-ALA-ASN-

  A3  |                    A4
LYS-ASN-LYS-GLY-ILE-THR-TRP-LYS-GLU-GLU-THR-LEU-MET-
                          CNBr1                    |

                       |
GLU-TYR-LEU-GLU-ASN-PRO-LYS-LYS-TYR-ILE-PRO-GLY-THR-
                           CNBr2

                           A5
LYS-MET-ILE-PHE-ALA-GLY-ILE-LYS-LYS-THR-GLU-ARG-GLU-
      |                          CNBr3

ASP-LEU-ILE-ALA-TYR-LEU-LYS-LYS--ALA-THR-ASN-GLU
```

Figure 7.4 Amino acid sequence of horse heart cytochrome *c* showing the expected cleavage sites for asparaginyl-endopeptidase (ASNase) as vertical lines above the sequence (the resulting peptide fragments are labeled A1 through A5), and the expected cleavage sites for cyanogen bromide (CNBr) as vertical lines below the sequence (the resulting peptide fragments are labeled CNBrl through CNBr3).

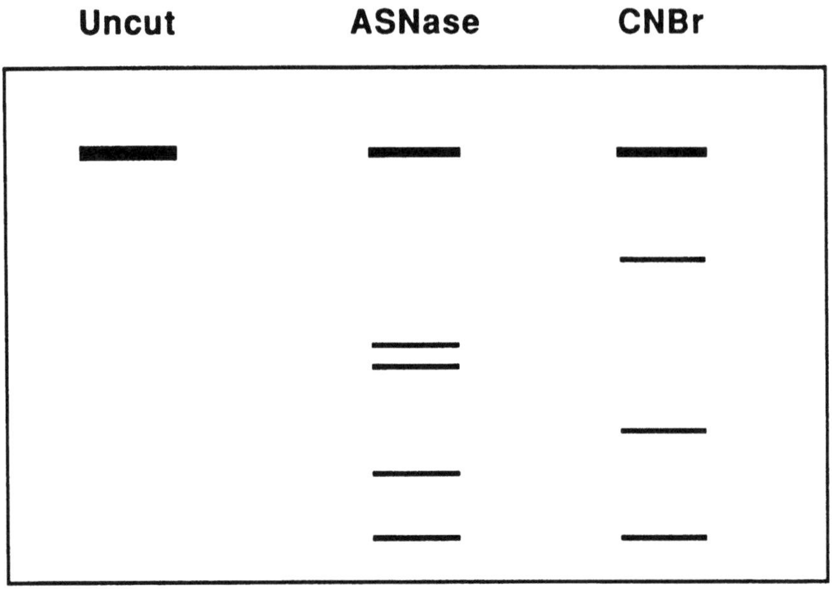

Figure 7.5 Cartoon of the expected results of SDS-PAGE for samples of horse heart cytochrome *c* before proteolysis (uncut) and after proteolysis with asparaginyl-endopeptidase (ASNase) and with cyanogen bromide (CNBr).

fication purposes. However, for a primary standard, or the first time one has purified a target protein, one may wish to further characterize the sample by determining its amino acid composition.

Proteins and peptides can be quantitatively decomposed to their constituent amino acids by several harsh chemical treatments that lead to peptide bond hydrolysis. By far the most common method for protein decomposition is acid hydrolysis (Darbre, 1986; Findlay and Geisow, 1989). It should be noted, however, that there are certain limitations to this method. First, one cannot distinguish Asn or Gln residues from Asp or Glu. Second, acid hydrolysis is destructive to tryptophan residues, and one cannot, therefore, quantitate these residues in this way. These limitations notwithstanding, acid hydrolysis remains the most widely used method for amino acid analysis of proteins and peptides, and we shall thus describe this method in detail here. Other methods for protein decomposition, such as alkaline hydrolysis, have been described by Darbe (1986) and by Hugli (1989).

Acid hydrolysis of proteins and peptides is performed at elevated temperatures (150° C) under extremely acidic conditions. Special glassware is thus required for this method. Protein hydrolysis tubes are available from Pierce and other manufacturers (see figure 7.6). The protocol described below for acid hydrolysis of proteins has been modified from the one recommended by Pierce.

MATERIALS

1. 6 N HCl (highest grade), 0.02% (v/v) 2-mercaptoethanol, 0.25% (w/v) phenol. Remove dissolved oxygen from this solution by bubbling nitrogen or argon through it for at least 5 min
2. Protein hydrolysis tubes with Teflon stoppers (Pierce)
3. Lyophilized or air-dried protein sample (1–5 mg)

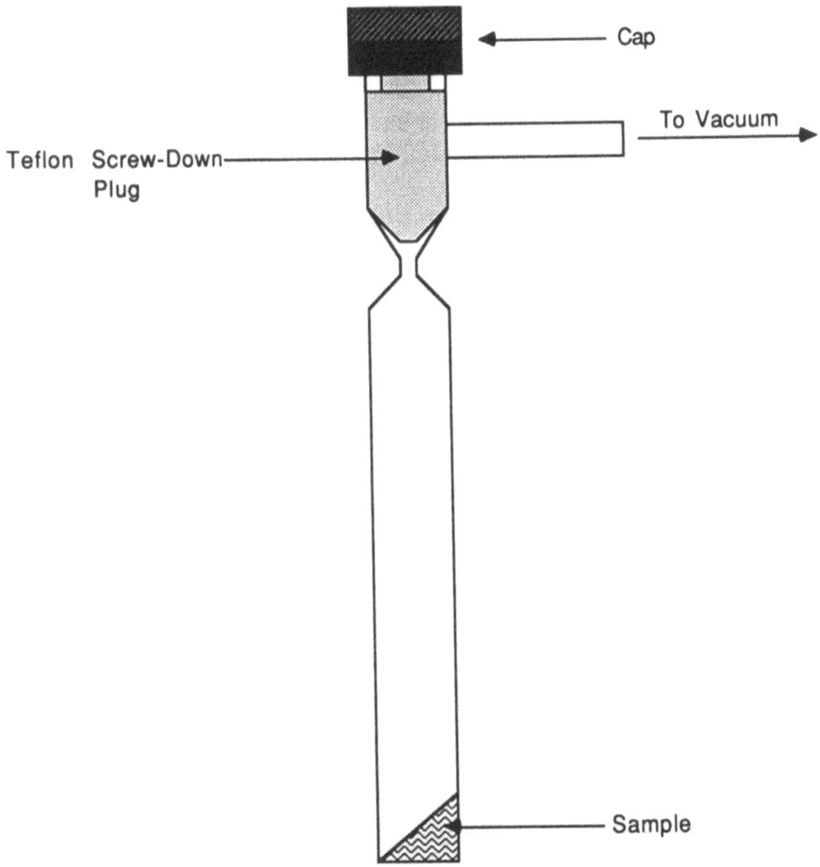

Figure 7.6 Diagram of a hydrolysis tube used for acid hydrolysis of proteins for amino acid composition analysis.

4. Vacuum and nitrogen sources
5. Alumimum block heater or oven capable of maintaining 150° C

PROCEDURE

1. Dissolve 1–5 mg of sample in 1 ml of 6 N HCl solution.
2. Remove the teflon stopper from a clean, dry hydrolysis tube, and place the sample solution into the bottom of the tube with a glass transfer pipette.
3. Insert the stopper and screw this down until there is only a small space between the plug and the glass at the constriction point, to allow gases to flow in and out of the tube. Connect the sidearm of the tube to a vacuum source, and evacuate the tube for a few min.
4. Close the stopper completely (do not overtighten) and disconnect the sidearm from the vacuum source. Attach the sidearm to a nitrogen source and slowly open the stopper until nitrogen flows into the tube. Close the stopper, and disconnect the sidearm from the nitrogen source. Repeat this cycle of vacuum and nitrogen application at least three times to ensure that all oxygen is removed from the tube. Apply a final vacuum for a few min, and close the stopper completely (again, do not overtighten).
5. Place the tube in an aluminum block heater (available from Pierce) or oven at 150° C for 6 h, or at 110° C for 20–72 h.
6. After hydrolysis, remove tube from heater and cool to room temperature. Connect sidearm to a nitrogen source, and slowly open stopper.
7. Remove the Teflon stopper from the tube. Using a glass transfer pipette, remove the sample and place in a small test tube, or other appropriate container, and lyophilize the sample as described in Chapter 2.

Detection of Hydrolyzed Amino Acids

There are numerous ways to separate and detect the amino acids derived from protein hydrolysis. Thin layer chromatography was used in the early days of protein chemistry for this purpose. Today, however, the most convenient method for separation, detection, and quantitation of amino acids involves covalent derivatization of the amino acids with a chromophoric group, followed by separation by reverse phase HPLC.

Phenylisothiocyanate (PITC) is commonly used to derivatize amino acids. PITC reacts with the amino group of amino acids (or the amino terminal residue of a protein) to yield the phenylthiocarbamyl (PTC) derivative of the amino acid. The PTC-labeled amino acid gives rise to a strong absorption band at 254 nm, providing a convenient means of detection. The molar absorbance for the PTC derivatives of the different amino acids have been determined for specific solvent systems (Heinrikson and Meredith, 1984). However, the best method for quantitation of amino acid content is to derivatize known quantities of pure amino acids, and run these as standard using one's own column and solvent system. A method for derivatizing samples after protein hydrolysis is given below.

MATERIALS

1. PITC (Pierce)
2. Highest grade pyridine
3. Highest grade triethylamine
4. HPLC grade acetonitrile
5. HPLC grade methanol
6. HPLC grade distilled water
7. lyophilized sample from protein hydrolysis.

PROCEDURE

1. Prepare coupling buffer as follows: acetonitrile:pyridine:triethylamine:water (10:5:2:3).
2. In a small glass test tube, dissolve the lyophilized sample in 100 μl of coupling buffer.
3. Place the tube in a vacuum flask and apply vacuum until all of the solvent is removed (make sure sample is completely dry). This step removes residual HCl from the sample.
4. Dissolve the dried sample in another 100μl of coupling buffer. Add 5 μl of PITC and incubate at room temperature for 5 min.
5. Dry the sample completely under vacuum, as in step 3.
6. Redissolve the sample in 250 μl of 7:2 (v/v) water:acetonitrile.
7. Inject an aliquot of this (1–10 μl) onto a reverse phase column (see below).

A variety of columns and solvent systems have been suggested for separation of derivatized amino acids following protein hydrolysis. Heinrickson and Meredith (1984) have suggested the use of C8 and C18 columns (5 μm) for separation. These authors performed the HPLC at 52° C, using the following solvent systems:

SOLVENT A

50 mM ammonium acetate (aqueous) adjusted to pH 6.8 with phosphoric acid.

SOLVENT B

(i) 100 mM ammonium acetate, pH 6.8 in 50:50 (v/v) acetonitrile:water, or

(ii) 100 mM ammonium acetate, pH 6.8 in 80:20 (v/v) methanol:water, or

(iii) 100 mM ammonium acetate, pH 6.8 in 44:10:46 (v:v:v) acetonitrile:methanol:water

SOLVENT C

70:30 (v:v) acetonitrile:water.

The analytical programs used varied with each solvent system. Using solvent B(i), for example, these authors used the following program:

Table 7.2 Program for separation of PTC-derivatized amino acids. From Heinrickson and Meredith (1984).

Time (min.)	% A	% B	% C
0	100	0	0
15	85	15	0
30	50	50	0
30.1	0	0	100
40	100	0	0

Waters Chromatography (a division of Millipore) has developed a commercial kit based on the PITC derivatization of amino acids described here. Their kit, known as the Pico.Tag system, provides an integrated system for post-hydrolysis amino acid analysis that includes derivatization reagents, eluent buffers, a reverse phase column, and a column heater. This system has become widely used in industrial laboratories because of its convenience, rapid sample throughput, and reproducibility. The manufacturer claims that, using this system, as little as one picomole of an amino acid can be detected. Protein hydrolysate, from as little as 100 ng of protein, can be completely analyzed in 12 min with this system.

Amino Acid Sequencing

The most definitive means of assessing the identity of a protein is the determination of its complete amino acid sequence. Amino acid sequence analysis is today performed with automated sequenators, based on the chemistry and engineering developed by Pehr Edman and his coworkers in the 1950's (Edman, 1960). Advanced detection methods, such as the use of fast atom bombardment mass spectroscopy (FAB-MS) have recently been introduced that significantly enhance the sensitivity of the method (Hugli, 1989). All of the methods used for modern protein sequence analysis are highly specialized, and by and large are not the realm of the generalist. Laboratories that are devoted full time to sequence analysis are present in many academic and industrial institutions. Since these methods are so specialized, the readers of this text are more likely to submit samples to such laboratories for sequence analysis, rather than perform the analysis themselves. For this reason, we shall only briefly review the chemistry involved in protein sequence analysis here. The reader who wishes to explore this aspect of protein analysis in greater depth is referred to several excellent books devoted exclusively to this technique (Findlay and Geisow, 1989; Hugli, 1989, and references therein).

The chemistry used by automated sequencers dates back to the 1950's and was largely developed by Pehr Edman and his coworkers. Each cycle of sequencing is the result of the following four steps:

1. coupling
2. cleavage
3. PTH conversion
4. detection.

Coupling

The free amino terminus of a protein is reactive with phenylisothiocyanate (in *n*-heptate) in the presence of base (usually trimethylamine). The reaction is done at 45° C under argon, and produces a phenylthiocarbamyl-derivative of the amino terminal residue of the protein (PTC-amino acid). Byproducts are removed by washing with *n*-heptane and ethyl acetate. PTC reacts exclusively with the free amino groups of amino acids, and the reaction is the same for all amino acids. Within a protein, the N-terminal amino acid is the only strongly reactive site. Weaker reactions can occur at the side chain amino groups of lysine, but these reactions are sufficiently slower than that at the amino terminus so that they do not present a major complication.

Cleavage

The PTC-derivatized amino terminal residue renders the first peptide bond highly reactive toward cleavage with trifluoroacetic acid. The reaction produces the anilinothiazolinine (ATZ) form of the terminal amino acid and the N-1 chain length protein with a new N-terminal residue which can be coupled again (i.e., cycle 2). The cleaved ATZ amino acid is extracted with chlorobutane.

PTH Conversion

The ATZ amino acids are very unstable. Therefore, they are converted to the much more stable PTH (phenylthiohydantoin) amino acid forms. This is done by heating to 64° C for 20 min in the presence of aqueous TFA. The TFA is subsequently removed by evaporation and the sample is dissolved in an appropriate solvent system, such as water/acetonitrile for HPLC. PTH amino acids are stable and absorb light at 270 nm, thus facilitating their detection.

N-terminal sequencing is performed by repetitive cycles of the four reactions described here. Thus, in the first cycle, the N-terminal residue is cleaved and detected. In the second cycle, residue number 2, which is now the new N-terminus, is coupled, cleaved, converted, and detected. This iterative sequence is repeated for each subsequent amino acid in the protein. There are, however, practical limits on the number of cycles that one can perform in this way. A primary factor in reducing the number of cycles that can practically be performed is the chemical yields obtained during coupling and cleavage for each cycle. The side products that build up during successive cycles eventually mask the detection of the true cleavage product for latter cycles. For this and other reasons, 20 to 30 cycles is typically the limit of unambiguous sequence determination. For many purposes, determining the sequence of the N-terminal 20–30 amino acids is more than sufficient for identification purposes. When more complete sequence information is required, proteolytic digestion of the protein is commonly performed, and the sequences of individual fragments are then determined. By applying more than one protease, and observing patterns of overlapping sequences, one can build up the complete sequence of a protein in this way.

With modern instrumentation, utilizing the above described chemistry, protein samples as small as a few picomoles can be successfully analyzed. Of course, sample purity is critical if one is to obtain meaningful information from sequence analysis. Within recent years it has become possible to separate proteins by SDS-PAGE, electrophoretically transfer the particular protein band of interest to polyvinylidene difluoride (PVDF) membranes, and directly sequence from these membranes. Moos et al. (1988) and Matsudaira (1987) have described methods for high yield transfer of proteins to such membranes and subsequent sequencing for picomolar quantities of proteins. PVDF membranes are commercially available from several sources, including Millipore who markets it under the trade name Immobilon.

Some common problems that are encountered in protein sequence analysis relate to post-translational modifications of the protein (see Chapter 6). As described above, the chemistry related to initial coupling requires a free amino group on the N-terminal residue of the protein. If the N-terminus is blocked by acetylation or the presence of an N-terminal pyroglutamate, for example, initial coupling cannot occur. This results in a failure to detect an amino acid signal in the first or subsequent cycles. If one's protein is amino-terminally blocked, treatment with a protease to generate peptide fragments can allow sequence information to be obtained (*vide supra*), for at least some of the fragments. Glycosylation at Asn or Ser/Thr residues can result in low yields for these residues because of difficulties in extraction. This problem can often be corrected by chemically or enzymatically deglycosylating the protein prior to sequencing (see Chapter 6). Likewise, phosphorylation of Thr or Ser residues can complicate the interpretation of the sequence data. Methods for handling these and other situations are well described in the recent text by Hugli (1989). Additional information on methods for sample preparation for sequence analysis should be sought from the specific laboratory that will perform the analysis.

References

Andrews, A. T. (1986) *Electrophoresis: Theory, Techniques and Biochemical and Clinical Applications*, 2d ed., Oxford University Press, Oxford.

Beynon, R. J., and Bond, J. S. (1989) *Proteolytic Enzymes: A Practical Approach*, IRL Press, Oxford.

Brown, J. R., and Hartley, B. S. (1966) *Biochem. J.*, **101**, 214–228.

Creighton, T. E. (1974) *J. Mol. Biol.*, **87**, 603–624.

Darbre, A. (1986) *Practical Protein Biochemistry: A Handbook*, Wiley, New York.

Dayhoff, M. O. (1969) *Atlas of Protein Sequence and Structure,* Vol. 4, National Biomedical Research Foundation, Silver Spring, MD.

Edman, P. (1960) *Ann. N.Y. Acad. Sci.*, **88,** 602.

Findlay, J. B. C., and Geisow, M. J. (1989) *Protein Sequencing: A Practical Approach,* IRL Press, Oxford.

Flannery, A. V., Beynon, R. J., and Bond, J. S. (1989) in "Proteolytic Enzymes: A Practical Approach" (R. J. Beynon and J. S. Bond, Eds.) IRL Press, Oxford, pp. 145–162.

Heinrikson, R. L., and Meredith, S. C. (1984) *Analyt. Biochem.*, **136,** 65–74.

Hugli, T. E. (1989) *Techniques in Protein Chemistry,* Academic Press, San Diego.

Matsudaira, P. (1987) *J. Biol. Chem.*, **262,** 10035–10038.

Mihalyi, E. (1978) *Applications of Proteolytic Enzymes to Protein Structure Studies,* 2d Ed., CRC Press, Boca Raton, FL.

Moos, M., Jr.; Nguyen, N. Y.; and Liu, T.-Y. (1988) *J. Biol. Chem.*, **263,** 6005–6008.

Swank, R. T., and Munkres (1971) *Analyt. Biochem.* **39,** 462–477.

8

Residue-Specific Chemical Modification of Proteins

Over the past 50 years a variety of reagents have been found that covalently modify specific amino acid residues within proteins. Such compounds can be useful analytical tools for assessing the distribution of a particular amino acid type between the interior and surface of a folded protein molecule. The rationale for this type of analysis goes as follows. If there exist buried and surface exposed populations of a certain amino acid type within a protein, then treatment with a covalent modifying reagent under native conditions will only result in modification of the surface exposed residues. Those residues within the hydrophobic interior of the protein will not make contact with the modifying reagent, and will thus be protected from covalent modification. Comparison of the number of modification events per molecule of protein in the folded and unfolded (denatured) states can thus be used to describe the distribution of that amino acid type within the protein. This type of information is important in mapping out the surface topography of native proteins, and for assessing changes in surface exposure under different solution conditions, after effecting a conformational transition of a protein, or during stability studies.

Several texts devoted to chemical modification of proteins have appeared that describe in detail modification methods for most of the amino acids (Darbre, 1986; Glazer et al., 1985; Lundbald, 1991). Rather than attempt a comprehensive survey of modification methods for all amino acids, we shall describe methods for specific modification of a few representative groups. The general strategy is similar in each case. One wishes to modify specifically a particular amino acid type, with little or no non-specific modification of other residues. The modifying agent should form a stable, covalent attachment to the amino acid side chain, and excess reagent or byproducts should be easily removable. Finally, an accurate means of quantitating the extent of reagent incorporation into the protein should be available. Detection can take the form of radiolabeling, absorption detection, fluorescence detection, etc. Because of safety and convenience issues, we shall restrict attention here to methods based on the incorporation of a chromophoric modifying reagent, so that quantitation can be easily performed by conventional absorption spectroscopy.

MODIFICATION OF TYROSINE RESIDUES BY NITRATION

Reaction of tyrosine residues with tetranitromethane (TNM) results in the formation of 3-nitrotyrosine which displays a greatly lowered pK_a for the phenolic hydroxyl group (~ 7). The deprotonated form of 3-nitrotyrosine absorbs strongly at 428 nm, making detection of the modified residues simple (Sokolovsky et al., 1966). This reagent has been successfully used to modify tyrosine residue in a large number of proteins (Lundbald, 1991). The specific conditions used for modification (protein concentration, buffer conditions, and molar excess of reagent) vary for different proteins, and optimal conditions for a specific protein must be determined empirically. The following procedure, however, provides a general method that will serve as a good starting point for most proteins.

MATERIAL

1. 0.15 M TNM (from Sigma) in 95% ethanol
2. PBS or other appropriate buffer for one's protein
3. 0.1 M Tris buffer, pH 9.0
4. Apparatus for dialysis or a desalting column (see Chapter 2)
5. Cuvettes and spectrophotometer.

PROCEDURE

1. Dilute the protein to a concentration of 2–4 mg/ml with PBS (2.5 to 10 ml is a convenient working volume), and place the solution in a small beaker with a stir bar. Begin stirring at room temperature.
2. Add 20 μl of TNM per ml of protein solution while stirring. Add the TNM solution in 20 μl additions to the vortex of the stirring solution and allow at least a few seconds for mixing between additions. After

adding all of the TNM, incubate the solution for 1 h at room temperature.

3. Stop the reaction by adding 200 μl of 1 M mercaptoethanol.
4. Immediately remove unreactive TNM and the yellow nitroformate byproduct of reaction by running the protein down a desalting column, or by extensive dialysis. In either case the protein should be exchanged into 0.1 M Tris buffer, pH 9.0.
5. Measure the absorbance of the sample at 428 nm, and quantitate the amount of nitrotyrosine produced using an extinction coefficient of 4,200 $M^{-1}\ cm^{-1}$.

SOME NOTES ON THIS METHOD

1. The absorbance at 428 nm is due to the deprotonated form of 3-nitrotyrosine, and thus the extinction coefficient is greatly dependent on the solution pH. Care must be taken to ensure that the final pH is 9.0.
2. Caution must be exercised when using and storing TMN as it is a potential explosive. Follow the supplier's instructions for safe storage and handling.
3. An additional advantage of this method is that 3-nitrotyrosine is stable to acid hydrolysis. Thus, one can also quantitate the amount of modified and unmodified tyrosine residue by amino acid analysis (see Chapter 7).
4. An alternative method for modifying tyrosine residues is by iodination. This method is extremely popular because the availability of radiolabled iodine isotopes (particularly ^{125}I) allows for detection of labeled protein at very low concentrations. The drawbacks of iodination include label incorporation at residues other than tyrosine, and double labeling of individual tyrosines to form 3,5-diiodotyrosine. Thus, iodination is not the method of choice for quantitative assessment of surface tyrosines, but is an extremely useful means of incorporating a detectable label into proteins at low concentrations. A detailed description of methods of iodination of proteins is given by Roholt and Pressman (1972) and by Thorell and Johansson (1971).

MODIFICATION OF TRYPTOPHAN RESIDUES WITH 2-HYDROXY 5-NITROBENZYL BROMIDE

Tryptophan residues can be covalently modified with a number of reagents that all tend to also react with cysteine residues. Among these reagents, 2-hydroxy-5-nitrobenzyl bromide (HNBB) is particularly useful. This reagent, first used by Horton and Koshland (1972), preferentially modifies tryptophan residues and imparts a strong absorbance at 410 nm that makes detection convenient.

MATERIALS

1. HNBB (Sigma)
2. Dry acetone
3. 0.5 M potassium phosphate, pH 6.5
4. Dialysis materials or a desalting column
5. 0.1 M Tris, pH 10.5.

PROCEDURE

1. Dilute the protein with phosphate buffer (or other appropriate buffer for one's protein) to a concentration of 20–40 μM. Place in a foil wrapped beaker with a stir bar and begin stirring at room temperature.
2. Weigh out a 50–100 molar excess of HNBB and dissolve in a *minimum* volume of dry acetone. Add this to the protein in several small volumes while stirring, and incubate *in the dark* for 2 h. Adjust pH to 7.0–7.5 as needed with dilute NaOH.
3. Remove excess reagent and byproduct by dialysis or gel filtration chromatography against Tris buffer, pH 10.5.
4. Quantitate HNBB addition spectrophotometrically at 410 nm using an extinction coefficient of 18.0 mM^{-1} cm^{-1}.

NOTES

1. As with the tyrosine nitration procedure, the pH is critical for accurate quantitation.
2. HNBB is photosensitive, so it is important to work in the dark. The reaction vessel should be covered with aluminum foil and, if possible, the room lights should be shut off during reaction.

MODIFICATION OF LYSINE RESIDUES WITH PYRIDOXAL PHOSPHATE

Lysine residues will react with pyridoxal-5′-phosphate (PLP) to form a Schiff's base complex. This linkage is labile, however, unless the Schiff's base is reduced by treatment with sodium borohydride. The bound PLP is both chromophoric and fluorescent, making it easy to quantitate its presence (Fisher et al., 1963).

MATERIALS

1. Pyridoxal-5′-phosphate (Sigma)
2. 0.1 M potassium phosphate, pH 6.0
3. Glacial acetic acid
4. Sodium borohydride
5. Dialysis materials or a desalting column.

PROCEDURE

1. Place the protein in a beaker with a stir bar and begin stirring.
2. While stirring add a 50–100 molar excess of PLP from a stock solution prepared in phosphate buffer. Incubate for 1 h at room temperature.
3. Adjust pH to 5.0 with acetic acid. Then add, while stirring, 50 mM sodium borohydride until ~ a 100 fold excess has been added.
4. Remove excess reagents by dialysis or other means (see above).
5. Quantitate PLP addition spectrophotometrically at 325 nm using an extinction coefficient of 9,710 M^{-1} cm^{-1} below pH 6.5 (Fisher et al., 1963).

MODIFICATION OF CYSTEINE RESIDUES WITH 4,4′-DITHIODIPYRIDINE

Grasetti and Murray (1967) introduced the use of 4,4′-dithiodipyridine (4-PDS) and 2,2′ dithiodipyridine (2-PDS) as chromophoric reagents for the quantiation of free sulfhydryl groups in protein. The 4,4′-dithiodipyridine (sold by Aldrich under the tradename of Aldrithiol-4) is the more stable of the two reagents, and displays little absorbance in the 280 nm region, where one might wish to assess protein absorbance. For these reasons, 4-PDS is more commonly employed than 2-PDS. This reagent reacts selectively with free cysteine residues within proteins, and gives rise to the byproduct 4-thiopyridone, which absorbs strongly at 324 nm. The byproduct is produced in amounts that are equivalent to the molar concentration of reactive sulfhydryls. Thus, if one knows the concentration of protein used, one can determine the number of reactive cysteines per protein molecule.

MATERIALS

1. 4-PDS (Aldrich Chemical Co.)
2. PBS or other appropriate buffer for one's protein
3. Protein stock solution
4. *Quartz* cuvettes and a spectrophotometer.

PROCEDURE

1. Prepare a stock solution of 4-PDS in buffer by dissolving a few grains of solid 4-PDS in 50 ml of buffer. Filter out any undissolved material, and determine the concentration of the reagent by measuring the absorbance at 247 nm using an extinction coefficient of 16,300 M^{-1} cm^{-1}. A convenient concentration for the stock solution is 50–100 μM. Use the reagent within the day or freeze it and store at $-20°$ C (the frozen solution is stable for at least a week).
2. To 3 ml of the 4-PDS solution add a sample of protein so that the final solution represents a 50–100 fold excess of 4-PDS over cysteine residues.
3. Incubate the solution at room temperature for 30 min (see notes).
4. Add to 3 ml of 4-PDS solution a volume of buffer equivalent to the volume of protein added in step 2. Place this in a quartz cuvette and use this solution to record a baseline for the spectrophotometer.
5. Record the absorbance at 324 nm for the solution from step 2. Quantitate the concentration of reactive sulfhydryls using an extinction coefficient of 19,800 M^{-1} cm^{-1}.

NOTES

1. Different proteins display different kinetics of reaction with 4-PDS. Generally 30 min is sufficient to ensure that the reaction has gone to completion, but it is worthwhile monitoring the kinetics of reaction for one's particular protein, at least for the first time the analysis is performed. This can easily be accomplished by setting the spectrophotometer at 324 nm, and monitoring the absorbance as a function of time. Figure 8.1 illustrates the type of time course one might expect to see. In subsequent experiments one would then use an incubation time that corresponds to the flat portion of the kinetic trace.
2. Another advantage of 4-PDS is that it, and its byproduct 4-thiopyridone, show little absorbance at 280 nm. One can therefore use absorbance at 280 nm as an internal monitor of the final protein concentration in one's sample. For comparative studies, it is often convenient

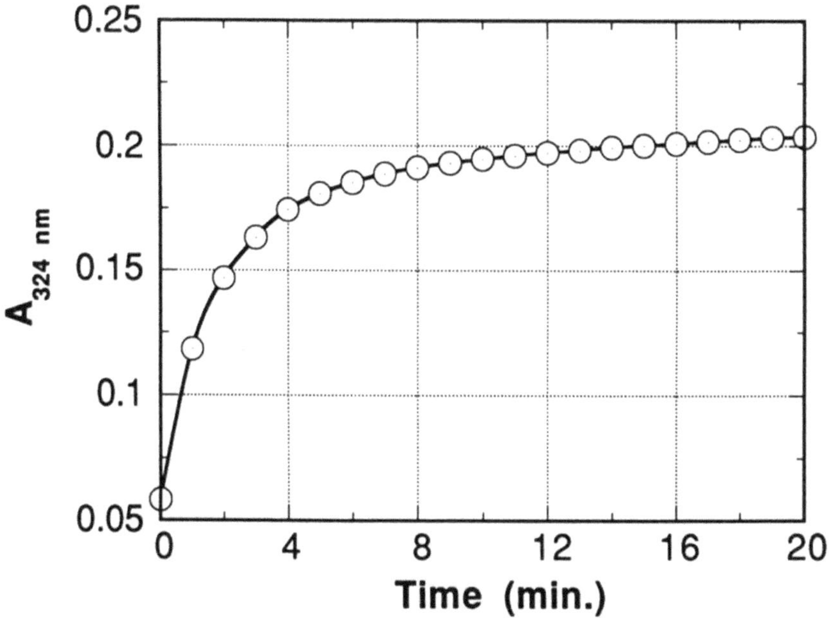

Figure 8.1 Time course of the reaction of bovine serum albumin with the cysteine-specific modifying reagent 4,4′-dithiodipyridine.

to report results in terms of the ratio of absorbance A_{324}/A_{280} to correct for any slight variations in protein concentration from one sample to the next.

3. Aldrich provides an excellent review on the use of this and other specific reagents for thiol group modification [reprinted from *Aldrichimica Acta* (1971) **4,** 33–48].

The procedures described above are illustrative of the methods generally used for amino acid specific modification of proteins. We have detailed methods for modification of four types of residues here. Detailed protocols for modification of other amino acids can be found in the references listed at the end of this chapter (Glazer et al., 1985; Lundbald, 1991).

As stated earlier, one of the more common uses of chemical modification is to probe the relative accessibility of certain amino acids within a protein under different conditions. For example, one could use chemical modification studies as a means of assessing the extent of protein unfolding that occurs under different conditions. Suppose that an amino

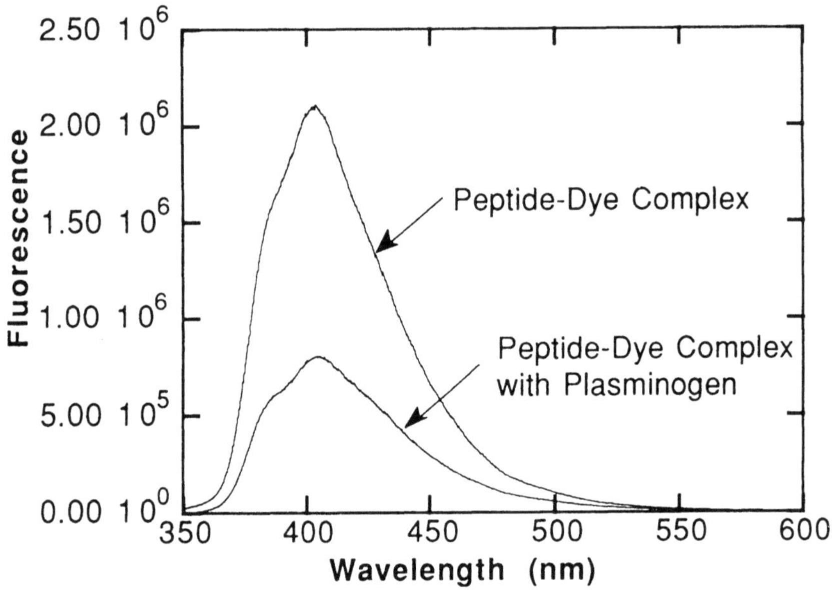

Figure 8.2 Fluorescence spectra of a fluorescently labeled tripeptide (Lys-Cys-Lys) in the presence and absence of the kringle-containing protein plasminogen. Data from Balciunas et al. (1993).

acid residue was buried in the interior of a protein when the protein was properly folded. In the folded state this residue would be inaccessible to covalent modifying reagents, and thus non-reactive. The reactivity of the residue would increase greatly under conditions that lead to unfolding of the protein, and hence exposure of the residue to the modifying reagent. This strategy was used, for example, by Liang and Terwilliger (1991) to study the unfolding of the gene V protein of bacteriophage f1. This protein contains a single cysteine residue that is buried within the interior of the folded protein. Liang and Terwilliger (1991) used the reactivity of this residue with 4,4′-dithiodipyridine as a measure of the degree of unfolding of the protein in solutions containing varying amounts of guanidine hydrochloride. As the fraction of unfolding protein increased with higher guanidine hydrochloride concentration, both the extent and rate of reactivity of the cysteine increased. This same group went on to use this method as a means of accessing the relative stability of mutants of the gene V protein compared to that of the wild type protein.

Another use of covalent modification reagents is to tag a protein with a reagent that imparts a distinct physico-chemical property to the molecule for selective detection of that protein in a mixture of reagents. We have mentioned, for example, the use of tyrosine iodination for attaching a radioactive label onto a protein. This method has been widely used to detect binding of radiolabeled proteins to cell surface receptors, and in other venues where direct detection of the protein is somehow constrained. An alternative to radiolabeling that has recently gained wide use is to attach a fluorescent molecule to the protein through covalent modification of specific amino acids. A wide variety of fluorescent modifying reagents are available for this purpose from Molecular Probes (Eugene, Oregon), with greatly varying spectroscopic properties. For instance, my own group recently exploited the use of a fluorescent modifying reagent to study peptide binding to the kringle domains of serum lipoproteins (Balciunas et al., 1993). Kringles are structural domains found in certain proteins that are thought to serve as binding sites for lysine residues on other target proteins. We synthesized a tripeptide containing lysine residues, Lys-Cys-Lys, that should bind tightly to kringles. The cysteine residue was included here as a site for specific covalent

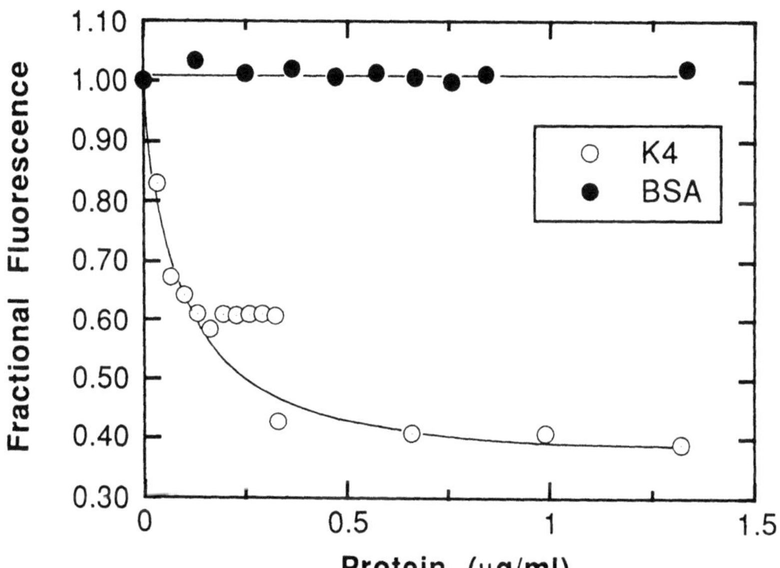

Figure 8.3 Fluorescence titration of a fluorescently labeled tripeptide (Lys-Cys-Lys) with proteins that contain the kringle lysine binding motif (K4) and that do not contain such a motif (BSA), illustrating the specificity of fluorescence quenching due to peptide binding to proteins containing kringle structures. Data from Balciunas et al. (1993).

modification with the sulfhydryl-specific modifying reagent 4-acetamido-4′-maleimidylstilbene-2,2′-disulfonic acid. This reagent fluoresces strongly at 405 nm when covalently attached to a sulfhydryl group. When the fluorescently labeled tripeptide bound to a kringle-containing protein, however, this strong fluorescence was quenched, as shown in figure 8.2. The degree of fluorescence quenching was related to the amount of kringle-containing protein added and was specific for kringle-containing proteins, as shown in figure 8.3. These properties of the fluorescently labeled tripeptide make it a potentially useful tool for studying the relative affinities of different kringle-containing proteins for lysine-containing peptides. Fluorescently labeled peptides and proteins have been widely used for studying protein-protein, protein-ligand, and peptide-receptor interactions in a variety of biological systems (for a review see Lackowicz, 1983).

References

Balciunas, A.; Fless, G.; Scanu, A.; and Copeland, R. A. (1993) *J. Protein Chem.*, **12,** 39–43.

Darbre, A. (1986) *Practical Protein Biochemistry: A Handbook,* Wiley, New York.

Fisher, E. H. A.; Forrey, W.; Hedrick, J. L.; Hughes, R. G.; Kent, A. B.; and Krebs, E. G. (1963) *Chemical and Biological Aspects of Pyridoxal Catalysis,* Pergamon Press, New York.

Glazer, A. N.; Delange, R. J., and Sigman, D. S. (1985) *Chemical Modification of Proteins,* Elsevier, New York.

Grasetti, D. R., and Murray, J. F., Jr. (1967) *Arch. Biochem. Biophys.*, **119,** 41–49.

Horton, H. R., and Koshland, D. E., Jr. (1972) *Meth. Enzymol.*, **25B,** 468.

Lackowicz, J. R. (1983) Principles of Fluorescence Spectroscopy, Plenum press, New York.

Liang, H., and Terwilliger, T. C. (1991) *Biochemistry,* **30,** 2772–2782.

Lundbald, R. (1991) *Chemical Reagents for Protein Modification,* CRC Press, Boca Raton, FL.

Roholt, O. A., and Pressman, D. (1972) *Meth. Enzymol.* **25B,** 438.

Sokolovsky, M.; Riordan, J. F., and Vallee, B. L. (1966) *Biochemistry,* **5,** 3582–3589.

Thorell, J. I., and Johansson, B. G. (1971) *Biochim. Biophys. Acta,* **251,** 363.

9

Spectroscopic Probes of Protein Structure

With few exceptions, the goal of most protein purification efforts is to obtain a sample that is not only pure, but that also maintains the protein in its native (i.e., biologically active) conformation. The ability to describe the conformation of a protein in solution, and to relate changes in conformation with biological activity, is thus a major focus of protein science. The most detailed description of protein structures come from the determination of the complete three-dimensional arrangement of protein components in space, from x-ray crystallographic or nuclear magnetic resonance (NMR) studies. Despite their power, however, these methods are not without their attendant drawbacks. X-ray diffraction studies of proteins are dependent on obtaining protein crystals of sufficient size and quality to yield usable diffraction patterns. This can often be a time consuming, and not necessarily successful, undertaking. Even when high quality crystals are obtained, solving the structure from the resulting diffraction patterns is a laborious and time consuming effort. Add to this the fact that certain classes of proteins, such as integral membrane proteins, are inherently difficult to crystallize, and one soon realizes that x-ray crystallography, while an extremely powerful method, is not a panacea for protein structural problems. Multidimensional NMR spectroscopy likewise suffers from certain difficulties that restrict its utility. Perhaps the greatest limit to the use of NMR spectroscopy for solving protein structures is that the complexity of the multidimensional data is such that the size of a protein that can reasonably be solved is limited to about 100 amino acids or so. While significant efforts are currently being put forth to push up the size limit for NMR spectroscopy, at least for the present this method is limited to relatively small proteins.

How then can the generalist glean information on the conformation of a protein in an expeditious fashion? Fortunately, there are a number of simple spectroscopic methods that have been developed over the years for assessing different aspects of protein structure. While none of these methods alone gives the same level of structural detail as do x-ray crystallography or NMR spectroscopy, in combination these methods can provide a reasonable description of the solution conformation of most proteins.

Since all of the methods described in this chapter rely on specific interactions between components of the protein and light energy, we shall begin with a brief description of the types of interactions between matter and electromagnetic radiation that are important in the spectroscopy of proteins.

INTERACTIONS OF ELECTROMAGNETIC RADIATION WITH MATTER

Spectroscopy can be most broadly defined as the interaction of electromagnetic radiation with matter (see Cantor and Schimmel, 1980; Campbell and Dwek, 1984; and Freifelder, 1982 for reviews). Any electromagnetic wave propagating along the z-axis in a laboratory fixed coordinate system can be considered to be composed of two wave trains at right angles to each other. One of these wave trains represents the electrical component (E), and the other represents the magnetic component (H) of the radiation. The spectroscopic methods we shall discuss here are concerned with the interactions of matter with the electrical component of the radiation only (figure 9.1). Restricting our attention for the moment to this electrical component, we have represented this component in figure 9.1 as a wave train defined by the xz plane. However, in fact, an electromagnetic wave propagating in the z direction can have its electrical component oscillating in any direction perpendicular to the z-axis (figure 9.2).

If we were somehow able to select only that electrical component that does oscillate in the xz plane (by, for example, using a polarizing filter), the resulting radiation would be referred to as *plane polarized* to denote the fact that we are dealing now with an electrical component restricted to a specific plane. Plane polarized light can be effectively used for defining the direction of transition dipoles in molecules (as in polarized absorption spectroscopy) and for estimating the rotational freedom of

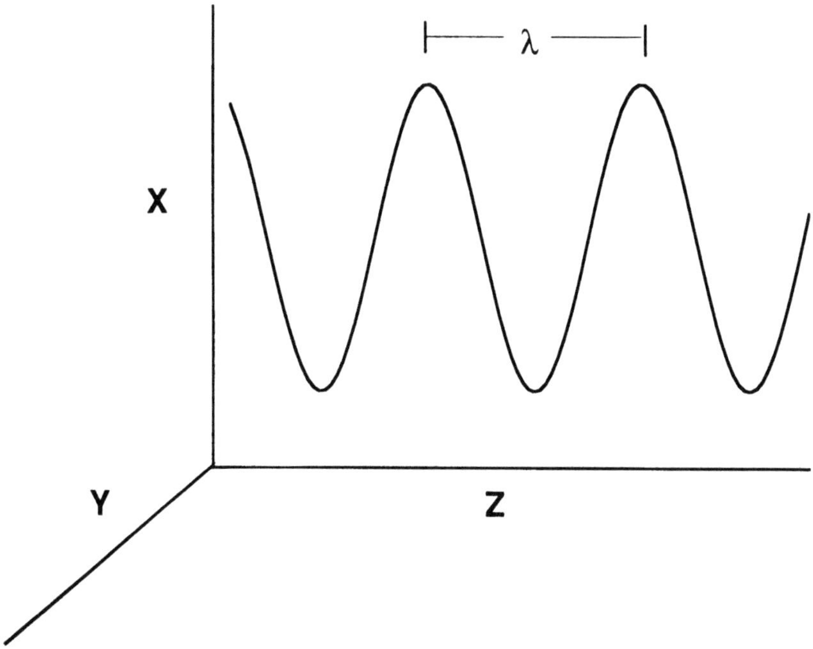

Figure 9.1 Diagram of the wave train associated with the electric component of electromagnetic radiation for radiation propagating along the z-axis. The wavelength (λ) is defined as the distance between two sequential wave crests.

fluorophores in proteins (through fluorescence depolarization measurements).

Now suppose that we had a radiation source with electrical components of equal intensity, oscillating in both the xz and yz planes. If these two wave trains were in phase with each other their resultant would be a new wave oscillating at 45° to the x-axis. If, however, these two waves were out of phase with one another by $\pi/2$, the resultant wave would not have a fixed axis of propagation, but would instead propagate as a helix. Such a resultant wave is said to be *circularly polarized*. Note that the helical propagation of such a wave can be clockwise (known as right circularly polarized) or counterclockwise (known as left circularly polarized). The use of circularly polarized radiation will become important in our discussion of circular dichroic spectroscopy.

The energy associated with any electromagnetic radiation is related to the frequency of the propagating wave train. The frequency is defined as the time that elapses between when a crest of the wave train passes a fixed point and when the next adjacent crest passes the same point.

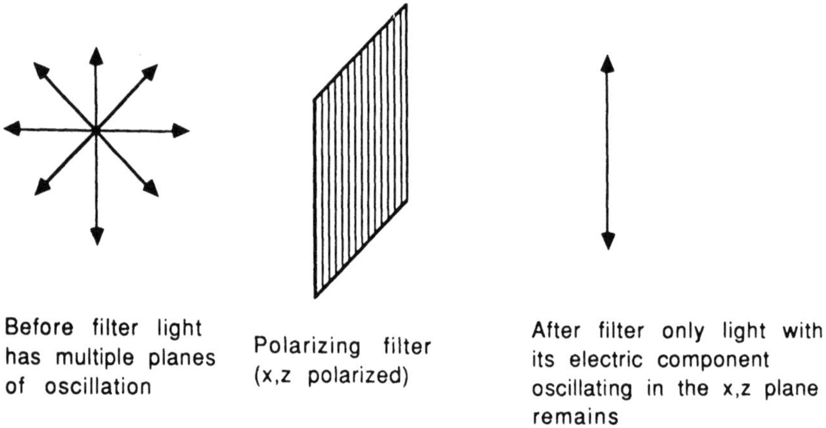

Figure 9.2 Polarization of light with a polarizing filter. Light propagating along the z-axis can have its electric component oscillating in any of multiple directions. After passing through a plane polarizing filter, only oscillations in one plane (i.e., xz) remain. Such light is referred to as plane polarized.

The wavelength of radiation is a convenient value for comparing different forms of radiation. This property, which is inversely related to energy, is defined as the distance between two adjacent crests of the wave train (figure 9.1). These properties are related as follows:

$$E = hc/\lambda = h\nu$$

where E is the energy of radiation, λ is the wavelength, ν is the frequency in hertz, h is Planck's constant (6.63×10^{-34}J•s), and c is the velocity of light (2.998×10^{8} m•s^{-1}). For convenience, the frequency is often converted from units of hertz to units of reciprocal centimeters (ν', cm^{-1}, or wavenumbers) by dividing by c:

$$\nu' = 1/\lambda = \nu/c$$

The wavelength is reported in units of length. For spectroscopic work the most convenient units used for wavelength are nanometers (nm; 10^{-9} meters). Angströms (Å; 10^{-10} meters), and microns (μ; 10^{-6} meters). In the older literature one finds spectroscopic data reported in all of these units. Today, however, most workers report data exclusively in nanometers.

The electromagnetic spectrum is conveniently divided into sections that correspond to different wavelengths of light (i.e., color) that, for historic reasons, are defined in terms of the sensitivity of the human eye in detecting these wavelengths. More importantly, these different sections of the electromagnetic spectrum correspond to the energies associated with discrete forms of molecular motions. Thus, as summarized in table 9.1, ultraviolet and visible light are of sufficient energy to invoke transitions among electronic states of a molecule, while infrared light can cause vibrational transitions, and microwave energy is sufficient to causes transitions among rotational states and electronic spin states of a molecule. It is important to bear in mind that the energies of these different types of molecular motions are such that each electronic state of a molecule will have associated with it a manifold of vibrational levels; each of these has an associated manifold of rotational levels, etc. If we diagram an electronic state of a molecule by the familiar Morse potential energy diagram, we can illustrate the above concept as shown in figure 9.3. In such a diagram, the y-axis (E) corresponds to potential energy, while the x-axis corresponds to interatomic distance.

For any type of molecular motion, the molecule has available a number of energetic states, the least energetic of which is referred to as the ground state. In the absence of any external energy, molecules will reside in their lowest energy (ground) states. What determines the efficiency with which light excitation will promote a transition from this ground state to a higher lying (excited) state is how well the energy of light is matched to the energetic difference between the two states involved. This concept is summarized in the Bohr frequency condition:

$$\Delta E = h\nu$$

where ΔE is the energy difference between the two states of the molecule, ν is the frequency of the exciting light, and h is Planck's constant.

Table 9.1 Regions of the electromagnetic spectrum and the molecular transitions that they induce.

Nature of Radiation	λ (nm)	Molecular Transitions Induced
Ultraviolet-Visible	2×10^2-7.5×10^2	Electronic
Infrared	10^6	Vibrational
Microwave	10^8	Rotational, and electron spin
Radio	10^{12}	Nuclear spin

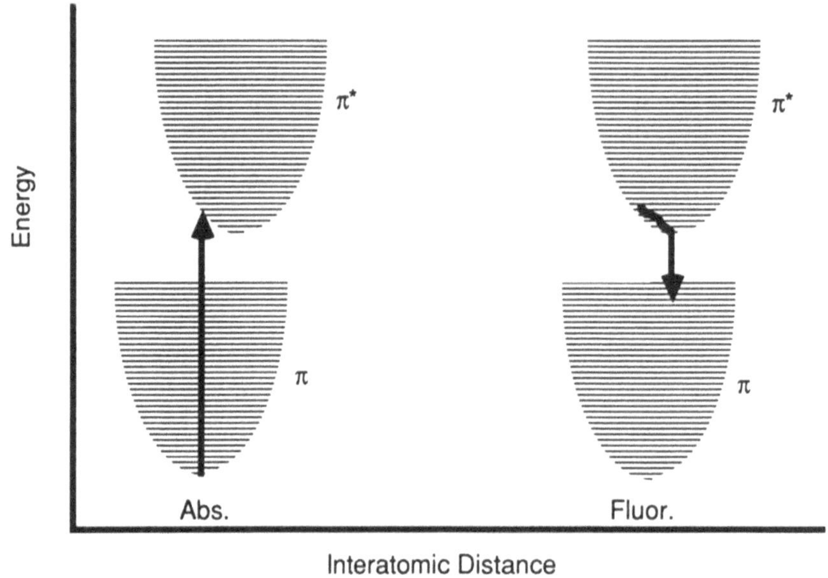

Figure 9.3 Energy level diagrams for absorption and fluorescence transitions between the ground state (π) and excited state (π^*) of a molecule.

Thus, only light of a very specific frequency will induce a molecular transition with high efficiency.

ULTRAVIOLET-VISIBLE ABSORPTION SPECTROSCOPY

Light in the ultraviolet (UV) and visible regions promotes transitions between electronic states of molecules. When restricting our attention to proteins and protein components, we are most concerned with moving an electron between π-type molecular orbitals within the molecule. This type of π–π^* transition is energetically diagrammed in figure 9.3. Note that the π^* state has more anti-bonding character than does the π state. Because of this, nuclei experience greater repulsion in the π^* state, and the potential minimum for this state thus occurs at a greater interatomic distance than that of the ground (π) state. An instantaneous transition from the ground state potential minimum to the π^* state would thus impinge on the excited state potential surface not at its minimum, but at a position corresponding to a higher lying vibrational level of the π^* state. If we label the vibrational sublevels of the π and π^* states as v_n, then, as drawn in figure 9.3, the most probable transition would be π, v_0 to π^*, v_4. This type of transition is referred to as a 0 to 4 transition. Note that one would also expect to observe 0 to 2, 0 to 3, 0 to 5, 0 to 6, etc., transitions, but with less probability. The end result is that one would expect a π–π^* transition to actually consist of a manifold of electronic-vibrational transitions. In practice, however, electronic transitions appear in the UV-visible spectrum as relatively broad, Gaussian band envelopes. Occasionally, however, one does observe evidence of the underlying vibrational fine structure built upon the electronic transition. A biologically relevant example of this is the UV spectrum of the amino acid phenylalanine, which shows remarkable resolution of its vibrational fine structure, as illustrated in figure 9.4.

BEER'S LAW AND SAMPLE ABSORPTION

The intensity of an absorption band is related to the concentration of absorbing molecules in the sample, the distance through which the light

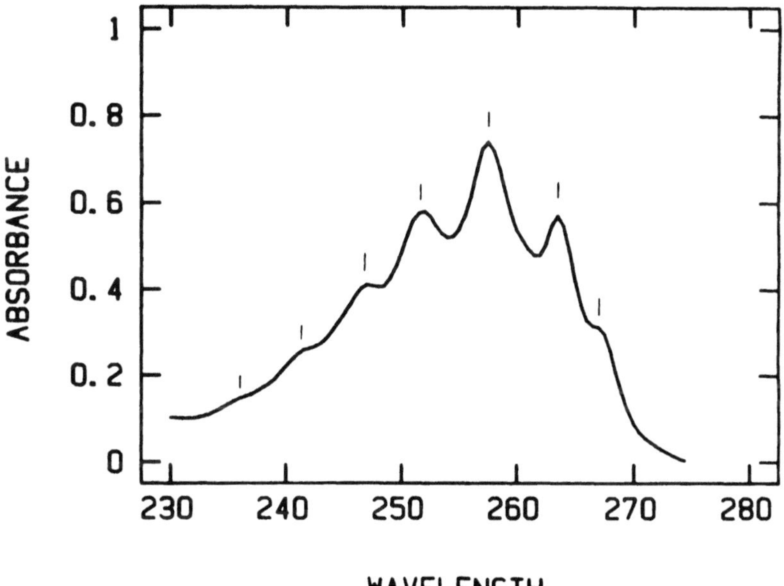

Figure 9.4 Absorption spectrum of the amino acid phenylalanine showing vibrational fine structure on the electronic transition envelope.

beam traverses the sample (i.e., pathlength), and an intrinsic property of the molecule itself, the extinction coefficient or molar absorptivity (ε). These are related by the following equation, known as Beer's law:

$$A = c\varepsilon l$$

where A is the absorption of the sample at a particular wavelength, c is the concentration of sample in molar units, ε is the extinction coefficient of the molecule, and l is the pathlength (usually measured in cm). Since absorption is a unitless quantity, the units of extinction coefficient must be reciprocal concentration x reciprocal length. Common units for extinction coefficients are M^{-1} cm^{-1}, and mM^{-1} cm^{-1}.

The form of Beer's law is that of a linear equation, so that one expects the absorption to increase linearly with sample concentration (assuming that the pathlength is kept constant). Indeed, one does observe the expected linear response within a certain range of sample concentration. As the concentration increases beyond this range, however, one begins to observe deviations from linearity, as seen in figure 9.5. Thus, to ensure the quantitative accuracy of one's data, it is important to work within the linear concentration range (referred to as the Beer's law limit). The point at which one begins to see deviations from linearity will depend on the sample and the instrument one is using. Some spectrophotometer manufacturers claim that their instrument will respond linearly for absorbance as high as 2.0. In general, however, it is best to work with sample absorptivities between 0.1 and 1.0. Below 0.1 signal to noise problems are often encountered, and above 1.0 one cannot be certain that the instrument will respond linearly.

One can also observe greater than expected absorption for samples that contain particles that scatter light. The light scattering in this case can lead to an apparent increase in absorptivity that can grossly distort the quantitative accuracy of one's measurement. To eliminate such problems it is imperative that all samples be filtered or centrifuged to remove protein aggregates and other particulates, prior to spectral data acquisition.

Another common source of error in absorption measurements has to do with the quality and cleanliness of cuvettes. As we have seen in previous chapters, many quantitative assays that rely on spectroscopic measurements can be accomplished with the use of plastic disposable cuvettes. It must be kept in mind, however, that any time one wishes to obtain accurate absorption readings below 350 nm, one must use high quality quartz (i.e., fused silica) cuvettes. Cuvettes must be rigorously clean before using them for spectroscopic measurements. A method for

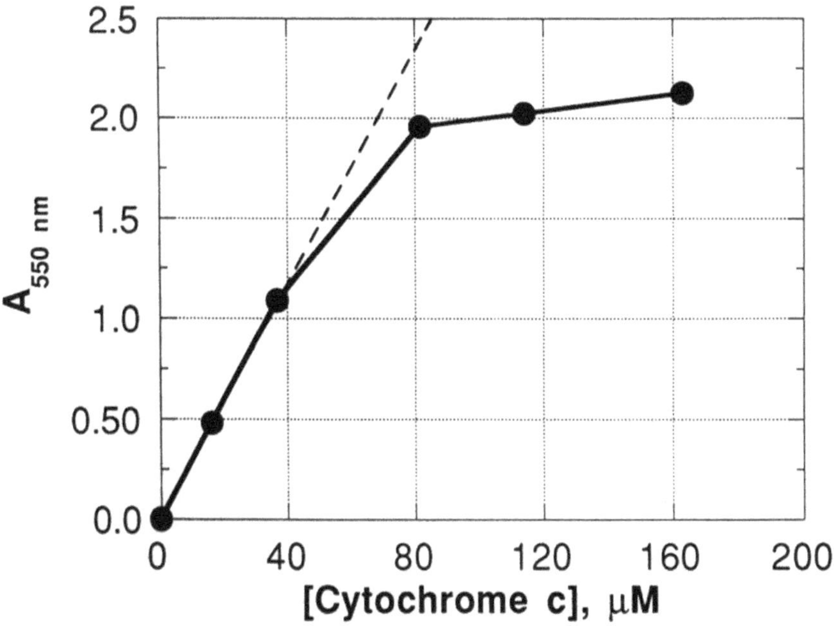

Figure 9.5 Apparent absorption at 550 nm for solutions of ferrocytochrome *c* as a function of protein concentration. Note that the expected linear response, based on Beer's law, is only valid over a limited range of concentration (up to an absorbance of 1).

cleaning quartz cuvettes for protein studies was given in Chapter 2. Never use any glass or metal instruments for cleaning cuvettes or for transferring liquids to and from cuvettes. The windows of cuvettes are high quality optical surfaces that can easily be distorted (scratched) by such utensils. Soft plastic transfer pipettes should always be used with cuvettes. If necessary, the surfaces of cuvettes can be scrubbed gently with a cotton-tipped swab, but usually the cleaning procedure described in Chapter 2 will suffice to clean these surfaces. Also, one should never touch the optical surface of a cuvette, as proteins and oils from one's fingers will absorb to the window and distort future spectroscopic data. Always wear gloves when handling optical cells.

ULTRAVIOLET ABSORPTION SPECTROSCOPY OF PROTEINS

The near ultraviolet region of the spectrum (250 to 350 nm) contains the lowest energy π–π^* transitions of the three aromatic amino acids, phenylalanine, tyrosine, and tryptophan (figure 9.6). All three of these residues derive their UV spectral characteristics from the substituted benzene rings that make up their side chains (Fodor et al., 1989). Phenylalanine has its maximum absorption at about 254 nm, and has a much lower extinction coefficient than the other aromatic amino acids (table 9.2). Thus, for most proteins, the contribution to this spectral region from phenylalanine is small, and can largely be ignored. The exceptions, of course, are proteins with unusually high numbers of phenylalanines in their sequences, or proteins that are devoid of other aromatic groups.

The other two aromatic amino acids, tyrosine and tryptophan, each give rise to a relatively strong π–π^* transition at 275 and 280 nm, respectively. As a result of the contributions from these two chromophores, most proteins display a composite absorption band envelope centered at about 278 nm in their UV spectra.

The utility of this composite absorption band in assessing protein structural changes is based largely on the sensitivity of the constituent π–π^* transitions to changes in solvent polarity and refractive index. Because of differences in electron distribution between the π and π^* electronic states, the energy of the π^* state is more strongly perturbed by changes in the polarity of the surrounding solvent. For aromatic molecules, like the amino acids discussed in this chapter, the net effect of this is that the π–π^* electronic transitions of these molecules are shifted to longer wavelengths (red shifted) when the solvent polarity is lowered. We can take advantage of this spectroscopic feature to assess the distribution of tyrosine and tryptophan residues within the protein interior and on the protein surface using several methods based on absorption spectroscopy.

Solvent Perturbation Spectroscopy

The first method that we shall describe, solvent perturbation spectroscopy, makes use of the fact that surface exposed residues will experience changes in environment when the solvent composition is changed. In the absence of any denaturation, however, residues that are buried within the interior of the protein molecule will not be affected by changes in solvent composition. Suppose, for example, that we were to supplement an aqueous buffer system with 20% ethylene glycol. This would have the effect of lowering the overall polarity of the solvent system in which the protein is solubilized. Aromatic residues on the surface of the protein would thus experience a lowering of the dielectric in their immediate surroundings; their resulting absorption spectra would be

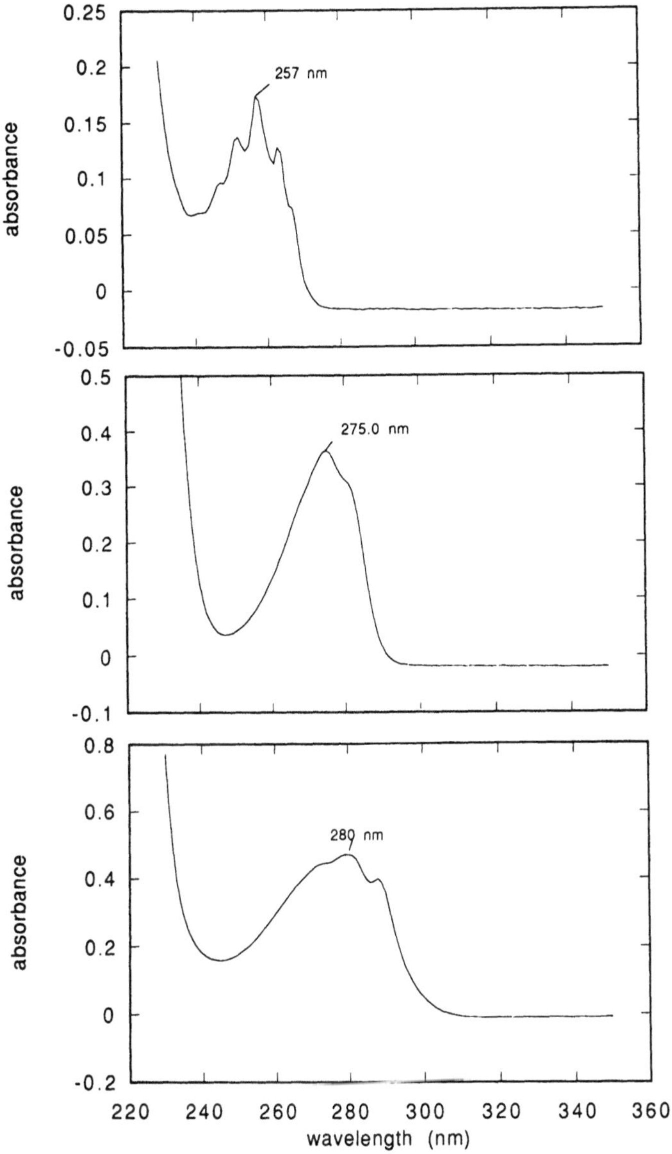

Figure 9.6 Absorption spectra for the three aromatic amino acids: phenylalanine (top, ~ 1 mM), tyrosine (middle, ~ 0.3 mM), and tryptophan (bottom, ~ 0.1 mM).

Table 9.2 Absorption properties of the aromatic amino acids.

Amino Acid	λ_{max} (nm)	ε ($mM^{-1}cm^{-1}$)
Phenylalanine	257	0.2
Tyrosine (pH 7)	275	1.4
Tyrosinate (pH 12)	294	2.3
Tryptophan	280	5.6

shifted in relation to their spectra prior to the addition of ethylene glycol (Herskovits and Sorensen, 1968; Donovan, 1973). The residues buried within the protein interior, on the other hand, would not experience this polarity change; the polarity of their immediate surroundings is defined by the protein structure itself. Therefore, if we measured the spectrum of the protein in buffer and again in buffer supplemented with 20% ethylene glycol, any differences we observed could be assigned to the residues on the exposed surface of the protein. The spectra themselves might show only minimal differences to the unaided eye. For example, addition of 20% ethylene glycol to aqueous solutions of *N*-acetyl tyrosine ethyl ester and *N*-acetyl tryptophan ethyl ester results in shifts of only 0.46 and 0.44 nm, respectively (Demchenko, 1986). These differences can be most clearly observed by taking the mathematical difference between the two spectra. Such difference spectra for tyrosine and tryptophan are shown in figure 9.7. In this figure each difference spectrum was computed by subtracting the spectrum of the sample in buffer from the spectrum of the sample in buffer plus ethylene glycol. The characteristic difference spectra seen for these amino acids are similar to what one would see for a solvent exposed residue on a protein surface upon solvent perturbation with ethylene glycol.

To put this method on a quantitative basis, one would need to know the wavelength maximum and difference extinction coefficient for the difference spectra. These values will vary depending on the specific perturbing solvent used. Fortunately, Herskovits and Sorensen (1968) have studied a large number of perturbing solvents and have calculated the wavelength maxima and difference extinction coefficients for their effects on *N*-acetyl tyrosine ethyl ester and *N*-acetyl tryptophan ethyl ester (the acetylation and esterification was done to eliminate charge effects that would not be encountered for residues within proteins). Their data is summarized in table 9.3.

We must also consider that in proteins we are likely to encounter a difference spectrum that reflects the sum of both tyrosine and tryptophan (and to a far lesser extent phenylanaine) differences. Because the difference spectral features for these two amino acids overlap, we need to consider the changes observed at more than one wavelength in order to quantitate the number of surface exposed residues of each type. Let us consider the case of using 20% ethylene glycol as our perturbing solvent. From table 9.3 we find that the wavelength maxima for tyrosine and tryptophan difference spectra, using this perturbant, are 286 and 292 nm, respectively. The table also provides us with the difference extinction coefficients for both amino acids at these two wavelengths. Following the method of Herskovits and Sorensen (1968) we can then set up two simultaneous equations in two unknowns as follows:

$$\Delta\varepsilon_{292} = (a)\ \Delta\varepsilon^{\mathrm{Trp}}{}_{292} + (b)\ \Delta\varepsilon^{\mathrm{Tyr}}{}_{292}$$

$$\Delta\varepsilon_{286} = (a)\ \Delta\varepsilon^{\mathrm{Trp}}{}_{286} + (b)\ \Delta\varepsilon^{\mathrm{Tyr}}{}_{286}$$

where *a* and *b* are the number of exposed tryptophan and tyrosine residues, respectively, $\Delta\varepsilon_{292}$ and $\Delta\varepsilon_{286}$ are the experimentally determined difference extinction coefficients at 292 and 286 nm, respectively, for one's protein, and the other difference extinction coefficients are ob-

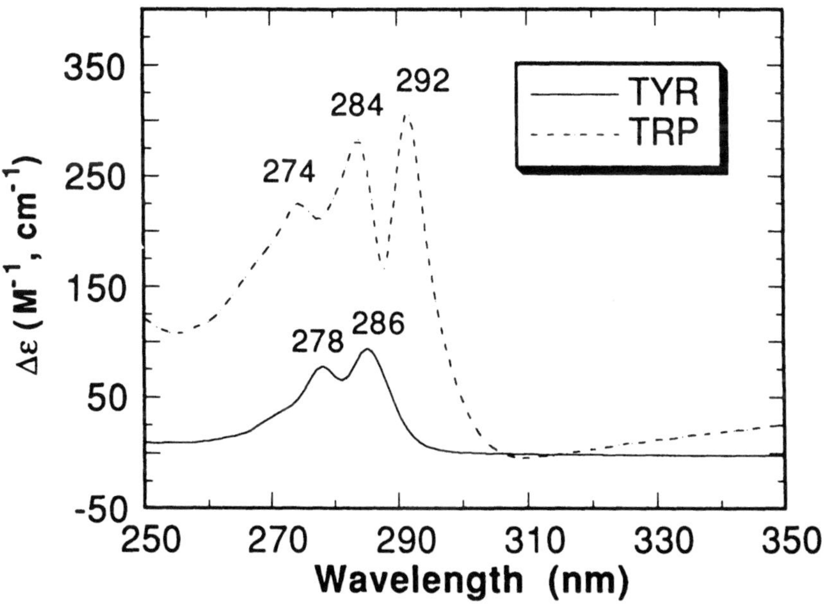

Figure 9.7 Solvent perturbation difference spectra for tryptophan and tyrosine induced by 20% ethylene glycol.

Table 9.3 Molar extinction coefficient different perturbants in aqueous solutions of N-acetyl tyrosine ethyl ester (Tyr) and N-acetyl tryptophan ethyl ester (Trp). Data taken from Herskovits and Sorensen (1968).

	Trp	$\Delta\varepsilon$ at Trp λ_{max}		Tyr	$\Delta\varepsilon$ at Tyr λ_{max}	
Perturbant	λ_{max}	TRP	TYR	λ_{max}	TRP	TYR
90% D_2O	292.0	−203.6	−12.2	285.5	−120.0	−67.1
20% Methanol	291.5	235.4	16.8	285.5	135.9	75.5
20% DMSO	292.5	489.5	35.5	286.0	168.4	213.7
20% Ethylene Glycol	292.0	305.1	16.1	285.5	172.2	92.1
20% Glycerol	292.0	304.4	12.9	285.0	195.6	79.8
20% Glucose	292.5	192.2	6.2	286.0	94.3	187.0
20% Sucrose	292.5	192.2	6.6	285.5	118.0	46.1

tained from table 9.3. Substituting in the values from the table we can recast these equations as:

$$\Delta\varepsilon_{292} = 305.1(a) + 16.1(b)$$

$$\Delta\varepsilon_{286} = 172.2(a) + 92.1(b)$$

If we multiply the second equation by 1.7718 and subtract the first equation from this, the resultant we obtain is:

$$(1.7718 \times \Delta\varepsilon_{286}) - \Delta\varepsilon_{292} = 147.1(b)$$

or:

$$[(1.7718 \times \Delta\varepsilon_{286}) - \Delta\varepsilon_{292}] / 147.1 = b = \text{number of exposed tyrosines}$$

Having thus solved for b, we can now go back into either of our original equations and solve for a. In this way we can simultaneously determine the number of exposed tyrosine and tryptophan residues in any protein sample. This same approach can, in principle, be used with any of the perturbants listed in table 9.3.

Folded-Unfolded Difference Spectra

When a protein unfolds, amino acid residues that were buried in the nonpolar interior of the protein become exposed to the polar aqueous solvent. Among these buried residues one often finds tyrosine and tryptophan groups. From the foregoing discussion one might think that the spectroscopic consequences of protein unfolding would thus be similar to what we have seen for solvent perturbation. The situation, however, is not quite so clear cut. In addition to the changes in solvent polarity that the buried residues experience upon protein unfolding, these residues also experience changes in environment due to the elimination of interactions with nearby amino acids, as well as changes in refractive index. These competing effects can alter the relative intensities of features in the difference spectra of different proteins. Nevertheless, the qualitative features of the difference spectra are similar to what we have seen for the aromatic amino acids themselves after perturbation of the solvent. For example, when bovine serum albumin is denatured with 6 M guanidine hydrochloride one observes a small blue shift of the absorption band maximum (figure 9.8).

Figure 9.9 illustrates the difference spectrum obtained by subtracting the spectrum of denatured BSA (in 6 M guanidine hydrochloride) from that of the native protein. For this protein the difference spectrum displays two positive features at 281 and 287 nm. From the wavelengths of these features we can surmise that both tyrosine and tryptophan residues

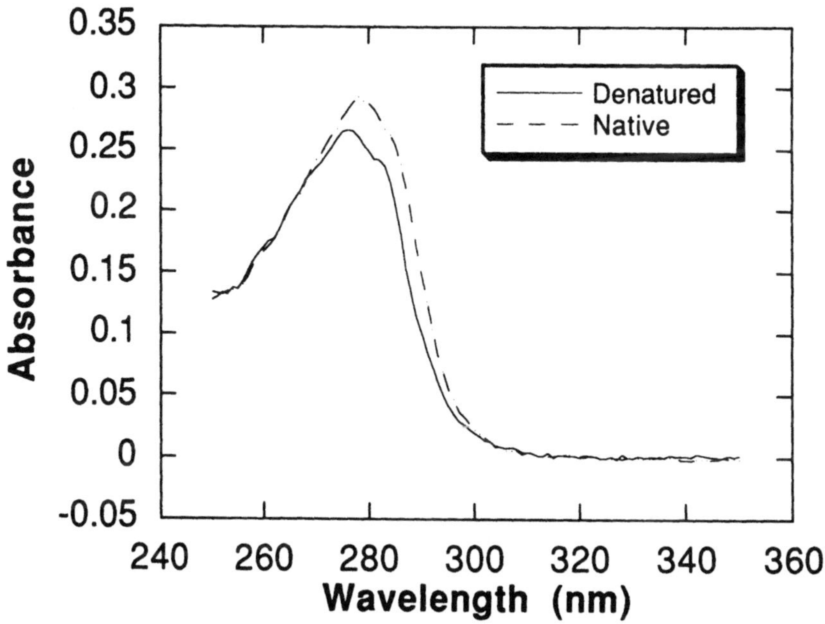

Figure 9.8 Absorption spectra of bovine serum albumin (BSA) in its native conformation and after denaturation of 6 M guanidine hydrochloride.

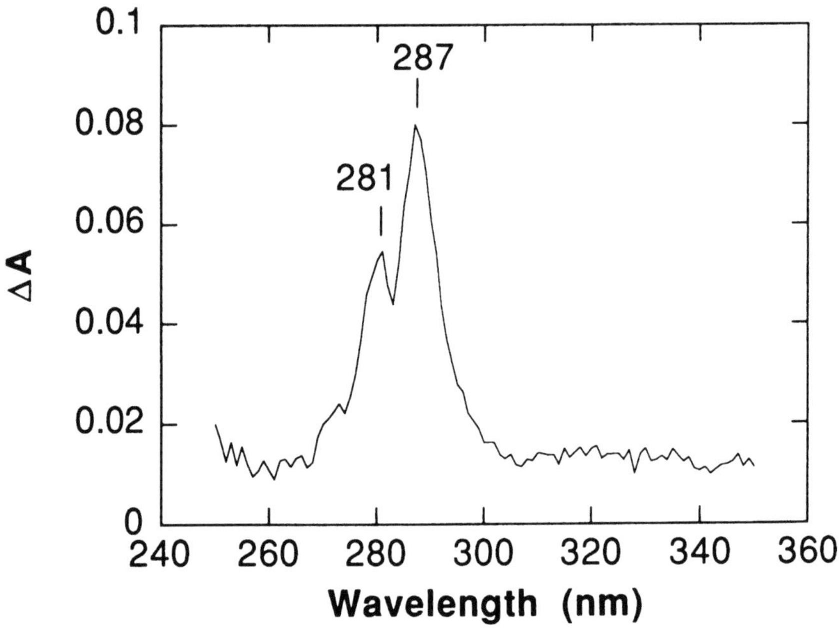

Figure 9.9 Difference spectrum of native minus denatured BSA calculated from the spectra shown in figure 9.8.

are experiencing changes upon protein denaturation. However, the largest contributor to this difference spectrum appears to be tyrosine, since no discrete band at ca. 292 nm is observed here. Thus, the pattern of bands in such a difference spectrum provides information on the types of aromatic groups that become exposed to solvent upon protein denaturation. While it is difficult to evaluate these data in a quantitative fashion, folded minus unfolded protein difference spectra are commonly used for following the kinetics of protein folding (and unfolding) as well as the equilibrium between these protein forms under different solution conditions.

At high concentrations of denaturant, the population of protein molecules in the denatured or unfolded state approaches 100%. The difference spectrum (relative to the native protein) obtained under such conditions would thus display the maximal changes one might observe. This is the situation illustrated in figure 9.9. What happens at lower concentrations of denaturant? Figure 9.10 illustrates the change in absorption at 287 nm that accompanies titration of BSA with guanidine hydrochloride. Note that at low concentrations of denaturant (below 2M for BSA) little protein unfolding takes place. At high concentrations of denaturant we reach a plateau representing completely unfolded protein (above 3.5 M for BSA). Between these extremes we have in solution a mixture of folded and unfolded protein. For a particular protein we can determine graphically the concentration of denaturant needed to achieve 50% unfolding (somewhere around 2.6 M for BSA). This value can then be used as a comparative measure of the inherent stability of proteins. In Chapter 10 we shall see how this type of experimental information can be used to evaluate protein stability under different conditions, and to compare the relative stabilities of different proteins.

The use of absorption differences between the folded and unfolded states of proteins as a means of following the kinetics of these processes is well documented in the literature. As an example, consider the difference spectrum for BSA that was shown in figure 9.9. From this spectrum we might choose to follow the rate of protein unfolding or folding by measuring the absorbance at 287 nm (the most intense feature in the difference spectrum). To follow protein unfolding what is typically done is to prepare a concentrated solution of protein under native conditions, and at the initiation of a kinetic run, to dilute this sample into a large

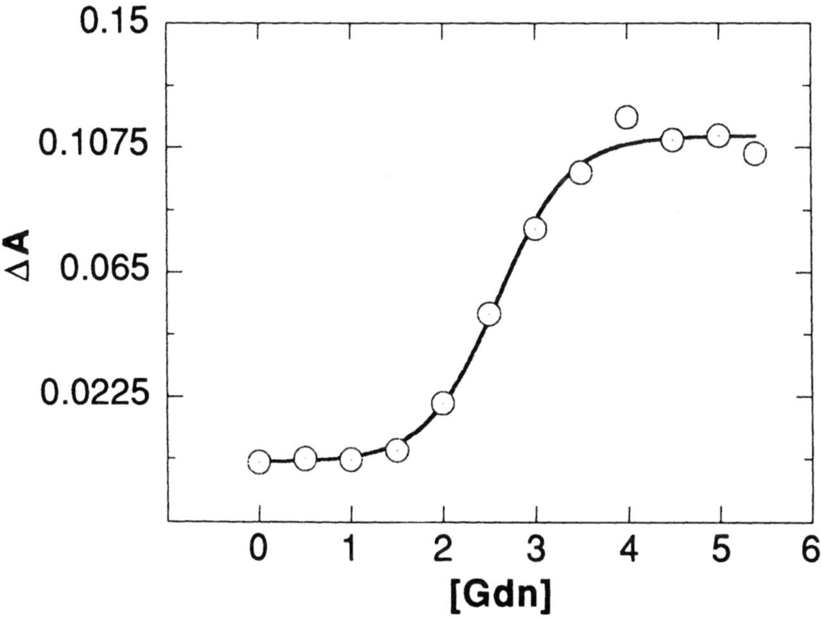

Figure 9.10 Titration of BSA with guanidine hydrochloride (Gdn) as measured by the absorption difference at 287 nm of samples minus the native protein spectrum.

volume of solution containing a high concentration of denaturant. As we shall see in Chapter 10, many proteins spontaneously fold into their proper three-dimensional structures in aqueous solution. For such a protein one could follow the refolding kinetics by reversing the experiment just described. That is, we would start with a concentrated solution of unfolded protein in a high concentration of denaturant and dilute this with buffer to lower the denaturant concentration to a point that would favor the folded protein state. As an example consider what we have learned about guanidine-induced unfolding of BSA from figure 9.10. Referring to this figure we see that at a denaturant concentration of 4.2 M we would expect nearly 100% unfolded protein in solution. If we then rapidly diluted the sample with buffer so that the final guanidine concentration was 1.62 M we would expect the protein to spontaneously refold. Figure 9.11 illustrates the results of exactly this type of experiment, in which the refolding of BSA was monitored by the time dependent increase in absorption at 287 nm. Data such as that presented in figure 9.11 are commonly used to determine the rate constants associated with protein folding and refolding under various solution conditions.

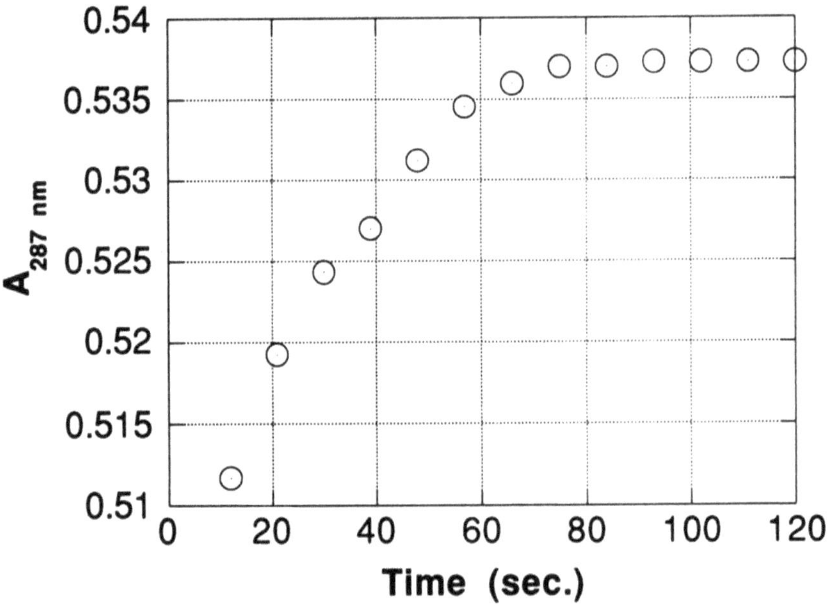

Figure 9.11 Kinetics of BSA refolding as monitored by absorption changes at 287 nm. At time zero the denatured protein was diluted with buffer to bring the Gdn concentration to 1.62 M. See text for further details.

PROTEIN FLUORESCENCE SPECTROSCOPY

We have seen how irradiation of a molecule with light of the correct frequency can elevate that molecule into a higher energy electronic state (excited state). From there the molecule will decay back to its lowest energy state, the ground state. Most times this relaxation process proceeds by the radiation of thermal energy (i.e., heat). In some instances, however, the molecule may return to the ground state by the emission of a photon (i.e., a light particle); for the π–π^* type of transitions we shall deal with here this light emission is referred to as fluorescence (Campbell and Dwek, 1984; Lackowicz, 1983).

Recall that when absorption of light leads to an instantaneous electronic transition, the excited state manifold is intercepted at a vibrational level higher than the lowest energy vibrational state. Before the molecule can return to the ground state by photon emission, it must first thermally decay to the lowest energy vibrational level of the excited electronic state. Because of this, the relevant energy gap between ground and excited state will be smaller for fluorescence than for absorption (see figure 9.3). This is manifested in the fluorescence spectral features always occurring at longer wavelengths (i.e., lower energy) than the corresponding absorption features; this difference in wavelength maxima for absorption and fluorescence is known as the Stokes shift. The size of the Stokes shift varies from molecule to molecule. For tryptophan, for example, the Stokes shift can be as large as 70 nm, depending on solution conditions (Lackowicz, 1983).

As with UV absorption spectroscopy, it is the aromatic amino acids that are most relevant to a discussion of protein fluorescence. Phenylalanine displays no appreciable fluorescence, and can thus be ignored for our purposes here. The two remaining aromatics, tyrosine and tryptophan, both fluoresce strongly when excited with UV light.

When excited at 275–280 nm tyrosine produces a fluorescence spectrum as shown in figure 9.12. The wavelength maximum for tyrosine occurs between 298 and 305 nm, but does not vary in any systematic fashion depending on environment. Although this amino acid displays relatively strong fluorescence, emission from these residues is rarely observed in proteins when tryptophan residues are also present. The protein backbone itself seems to quench (attenuate) the fluorescence of tyrosines. However, when no tryptophan is present in the amino acid sequence of a protein, tyrosine fluorescence is observed, albeit at weaker overall intensity than would be expected for an equimolar solution of the free amino acid. When tryptophan is present, however, virtually all of the tyrosine emission is quenched. The mechanism for this is not fully

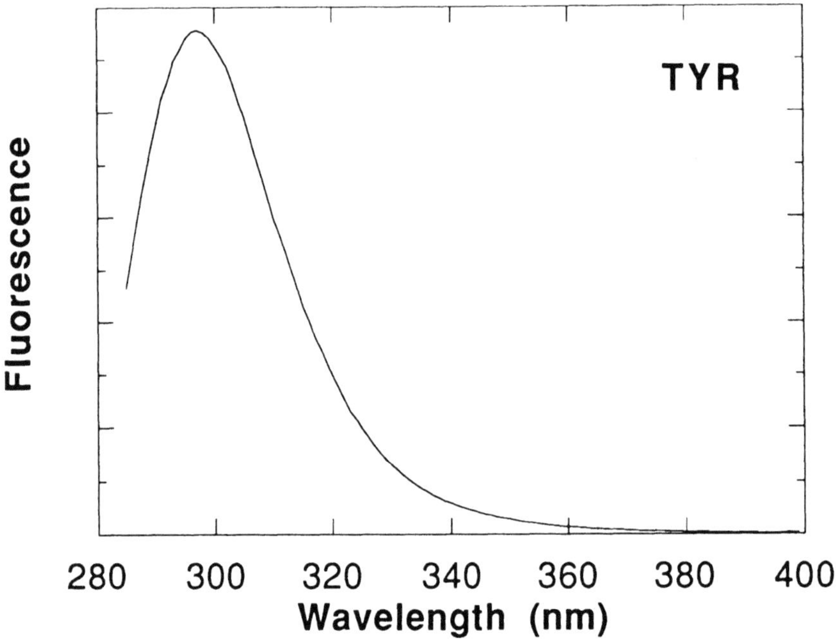

Figure 9.12 Fluorescence spectrum of tyrosine in PBS with excitation at 280 nm.

understood. In part this effect may be due to reabsorption of the tyrosine emitted photons by the tryptophan, but this alone would not explain the degree of quenching observed in proteins. Even when tyrosine is present in high molar excesses over tryptophan, one usually observes only weak signals from the tyrosine residues on top of the intense background of tryptophan fluorescence (Lackowicz, 1983).

Thus, the use of tyrosine fluorescence spectroscopy is usually limited to proteins that are devoid of tryptophan residues. In these cases one usually finds that the intensity of tyrosine fluorescence is a more useful marker of changes in environment than is the wavelength maximum of emission. For example, several DNA binding proteins are known that do not contain tryptophan in their amino acid sequences. Binding of DNA by these proteins leads to a dramatic quenching of tyrosine fluorescence due to π–π interactions between the tyrosine residues and the nucleic acid bases. This quenching is relieved when the complex between the two macromolecules is disrupted by increased ionic strength. These features of tyrosine fluorescence have made the technique a powerful method for determining the relative binding affinities of different nucleic acid polymers for particular proteins, and for assessing changes in the

DNA binding capacity of a protein that accompany point mutations of its amino acid sequence. For example, Zabin and Terwilliger (1991) used tyrosine fluorescence quenching to determine the relative binding affinity for DNA of mutants of the bacteriophage f1 gene V protein. They found, among other things, that replacement of Arginine 16 by a cysteine residue in this protein dramatically lowered the amount of salt required to reverse the DNA-induced tyrosine fluorescence quenching (Zabin, 1991). The positive charge on this arginine is thus likely to participate in ionic interactions with the phosphate backbone of the DNA which stabilize the complex.

While most tryptophan-containing proteins show minimal tyrosine fluorescence in their native conformations, there are a few exceptions to this general rule. A small number of proteins have been found that contain multiple tyrosines and a single tryptophan residue, yet they display mainly tyrosine fluorescence in their native forms. This group of proteins includes the mitogen acidic fibroblast growth factor (Copeland et al., 1991), and the visual protein arrestin (Kotake et al., 1991). In these cases denaturation of the protein leads to the appearance of tryptophan fluorescence. Therefore, the single tryptophan of these proteins must be in some special environment within the native protein that leads to significant quenching of the tryptophan fluorescence.

Tryptophan fluorescence is much more commonly observed in proteins than is tyrosine fluorescence, and has formed the basis for a wide variety of protein characterization methods. Unlike tyrosine, the wavelength maximum of tryptophan fluorescence is significantly affected by changes in solvent polarity. As the polarity of the solvent decreases, one observes a steady shift of the tryptophan fluorescence maximum to lower wavelengths. In water, for example, the fluorescence maximum for tryptophan occurs at 350 nm (figure 9.13), while in hexane it is shifted to below 310 nm (Lackowicz, 1983). The same trend is seen for tryptophan residues within proteins. When tryptophan resides within the nonpolar interior of a protein, its fluorescence maximum is typically seen at 325–330 nm. Tryptophan residues on the surface of a protein fluoresce maximally at 345–350 nm. For proteins containing a single tryptophan, the wavelength of maximal fluorescence is thus diagnostic of the environment that the residue experiences. For proteins with multiple tryptophan residues, the fluorescence wavelength maximum will reflect the sum of the fluorescence spectra of the individual residues. Even in the case of multiple tryptophan-containing proteins, however, fluorescence spectroscopy offers a very powerful tool for following changes in protein structure. Often times changes in conformation that accompany ligand binding, DNA interactions, protein unfolding, etc., will affect the envi-

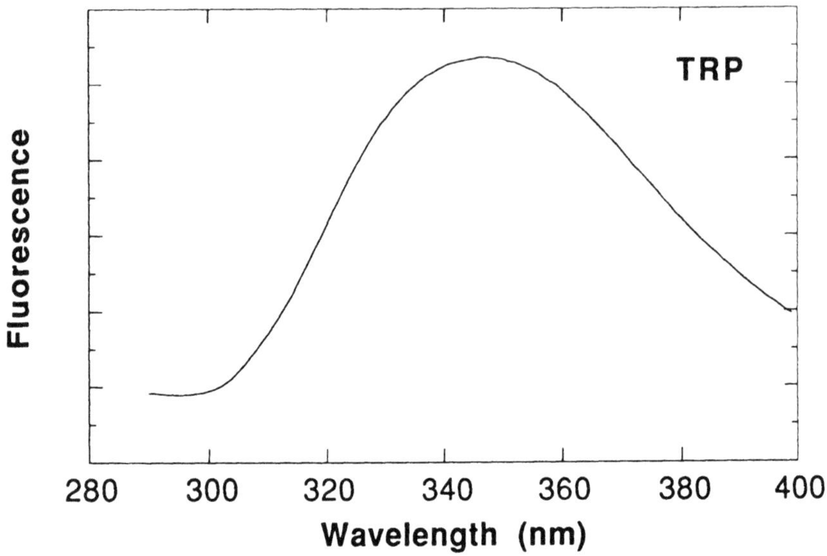

Figure 9.13 Fluorescence spectrum of tryptophan in PBS with excitation at 280 nm.

ronment of some or all of the tryptophan residues within the protein. Even when only a small subset of the tryptophan residues are affected by a conformational change, the fluorescence spectrum is often sensitive enough to report that change. Numerous examples of this can be found in the biochemistry literature from the past 30 years (see, for example, Chen and Edelhoch, 1975).

As an example of the sensitivity of tryptophan fluorescence to environment, consider the fluorescence spectra of folded and unfolded BSA, shown in figure 9.14. In its native (folded) state, the tryptophans of this protein are buried in the protein interior, and hence give rise to a fluorescence spectrum with a wavelength maximum at 335 nm. Despite a large excess of tyrosine residues over tryptophan in this protein, one does not detect an independent fluorescence signal from tyrosine here. When the protein is denatured with guanidine hydrochloride, the tryptophan fluorescence maximum shifts to the red, now occurring at 345 nm. Note that with unfolding, one now observes a tyrosine signal as a shoulder on the blue edge of the broader tryptophan emission envelope. In a number of proteins, tyrosine emission is observed after denaturation. This probably results from two effects. First, the shifting of the tryptophan fluorescence maximum to a longer wavelength tends to "uncover" any underlying tyrosine fluorescence that was masked in the spectrum

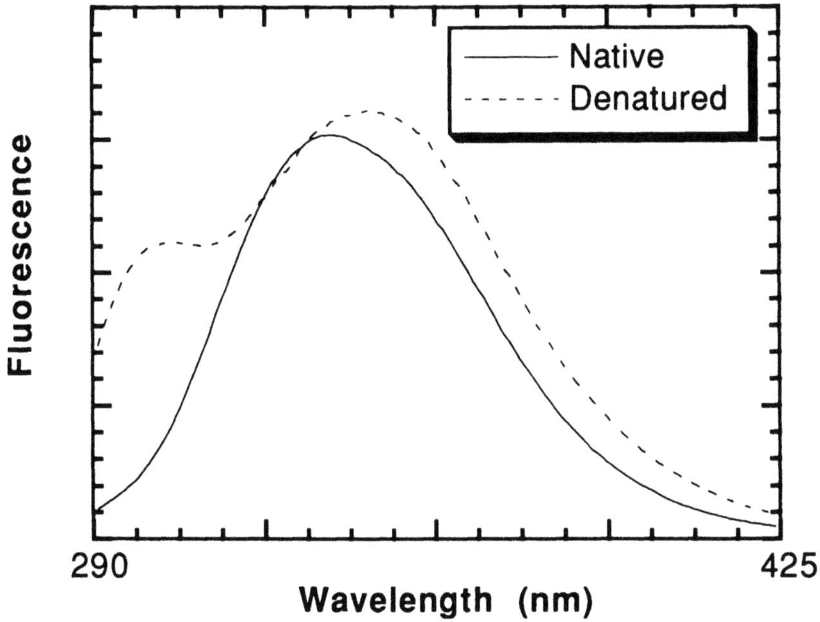

Figure 9.14 Fluorescence spectra of native and denatured (in 6 M Gdn) BSA with excitation at 280 nm.

of the folded protein. Second, it seems that the mechanisms that lead to quenching of tyrosine fluorescence in native proteins can be partially relieved in the denatured state.

Although tyrosine fluorescence is highly quenched in proteins, it can still contribute to the overall emission envelope, as we have seen in the spectrum of denatured BSA (figure 9.14). This raises the concern that for the native protein, the observed wavelength maximum for tryptophan fluorescence may be compromised by blue edge contributions from tyrosine residues, especially for proteins with high tyrosine content. To overcome this potential problem, one can take advantage of the differences in absorption spectra for tyrosine and tryptophan to find an excitation wavelength that selectively excites tryptophan. Comparing these absorption spectra (figure 9.6) one finds that at 295 nm, tryptophan still has moderate absorption while tyrosine does not. Thus, excitation at 295 nm will lead to tryptophan fluorescence with little, if any, contribution from tyrosine. The disadvantage of this method is that the fluorescence intensity, which is dependent on the absorption extinction coefficient at the excitation wavelength, will be greatly diminished at 295 nm relative to 280 nm excitation. For most proteins the tyrosine fluorescence contribution is minimal, even with 280 nm excitation. Even when one does need to resort to 295 nm excitation, one can usually achieve good signals with relatively low protein concentrations (see next section).

Corrections to Protein Fluorescence Spectra

Although obtaining a fluorescence spectrum of a protein is at first glance a trivial exercise, there are a number of potential interferences that can alter the data, and thus lead to misinterpretations. In this section we shall discuss some common problems encountered with fluorescence spectroscopy of proteins, and methods for avoiding or correcting for these problems.

The first problem to be discussed is that of non-protein signals from buffer components. There are three major types of interferences from buffer components: light scattering from particulates, Raman scattering from solvent, and fluorescence from non-protein components of the buffer. Light scattering is handled by filtration or centrifugation as previously described. Light scattering effects are particularly problematic in fluorescence spectroscopy, so that one must be especially careful to filter all sample solutions prior to spectral acquisition.

Raman scattering from solvent becomes a problem when one is working with very dilute protein solutions, where the overall fluorescence intensity is weak. Under these conditions one can observe a sharp feature in the fluorescence spectrum that occurs at a displacement from the excitation wavelength of ca. 3500 cm^{-1}, and is due to a Raman active vibrational mode of water (figure 9.15). With 280 nm excitation, this feature occurs at about 310 nm and can thus be mistaken for tyrosine emission. However, this vibrational band is extremely narrow compared to a true fluorescence signal. Also unlike a fluorescence signal, its wavelength maximum varies with excitation wavelength (i.e., it maintains a constant displacement from the excitation wavelength). Table 9.4 lists the wavelength maximum for the water Raman band at a number of excitation wavelengths that are commonly used for protein spectra. The simplest means of correcting for this solvent signal is by obtaining the fluorescence spectrum of a buffer blank under the exact conditions that one will use for the protein sample, and subtracting this blank from the subsequent spectra of one's samples. While solvent Raman scattering is usually only observed for very dilute protein solutions, it is so easily corrected for that subtraction of a buffer blank should be a routine part of one's data acquisition method for all protein samples.

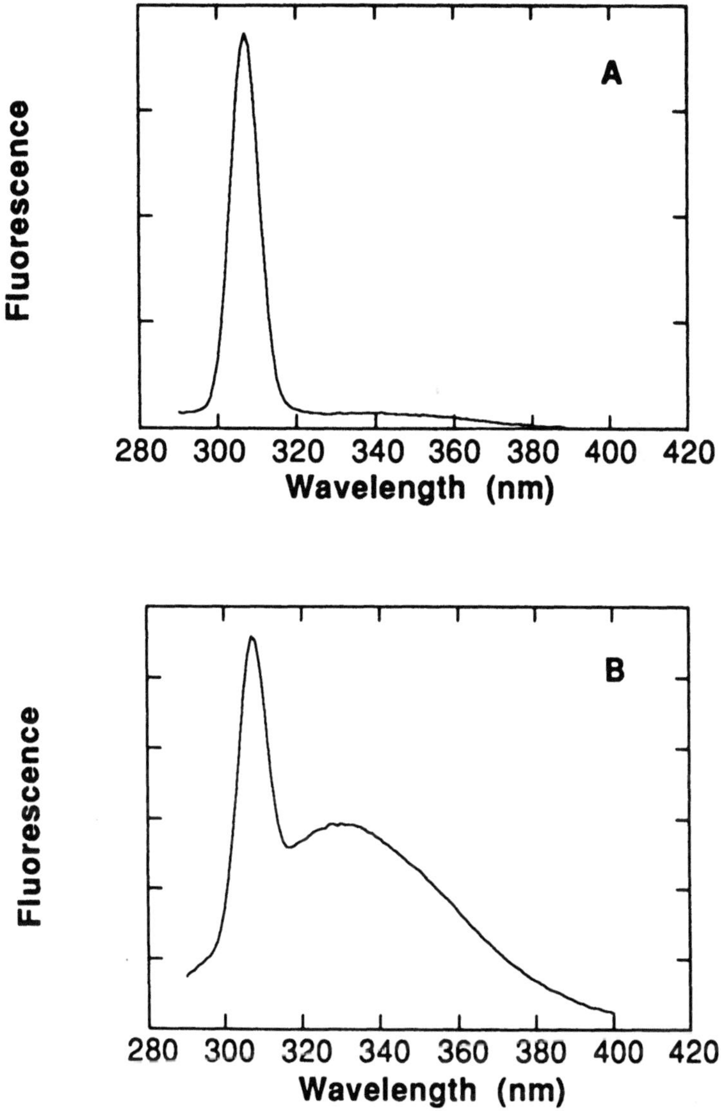

Figure 9.15 Raman scattering from water observed with a fluorimeter for samples of distilled water contained in a properly cleaned cuvette (A) and in a cuvette with residual protein adsorbed to its optical surfaces (B).

Table 9.4 Expected wavelength of the water Raman band at different excitation wavelengths.

$\lambda_{excitation}$ (nm)	λ_{Raman} (nm)
275.0	303.8
280.0	309.9
285.0	316.1
290.0	322.2
295.0	328.4
300.0	334.6

The final form of contributed signal comes from buffer components, other than the protein, that fluoresce. This is a relatively uncommon problem if one uses high quality reagents for preparing solutions. Occasionally, however, one will encounter a fluorescence signal from the buffer. In this case it is usually best to discard the buffer and remake the solutions. For membrane bound proteins, detergents offer an additional source of fluorescence. The Triton series of detergents should be avoided where possible because they often contain fluorescent impurities and have high UV absorption. Octyl glucoside and dodecyl maltoside are usually the best detergents from a spectroscopic point of view. Of course, one must take into consideration the effects of different detergents on the biological activity of the protein as well.

Another source of erroneous results comes from protein contamination on the optical surfaces of cuvettes. Figure 9.15 illustrates the fluorescence spectra of distilled water (Raman scattering) taken with a properly cleaned cuvette, and in a cuvette with a small amount of residual protein adhered to its optical windows. Such contamination can result from insufficient cleaning of the cuvette interior after use, or from touching the outer optical surface with one's hands. Proper procedures for handling and cleaning cuvettes have been described earlier (Chapter 2). These procedures should always be followed.

The intensity of fluorescence is directly related to the absorption extinction coefficient of the molecule at the excitation wavelength. One would thus expect that the fluorescence intensity would increase with increasing sample absorption (i.e., concentration). However, if the absorption of the sample is too high, light emitted by molecules in the sample may be reabsorbed by other molecules, leading to a diminished apparent molar quantum yield (i.e., emission intensity). This so called inner filter effect becomes critical when sample absorption is greater than ca. 0.1. In general, one should use samples whose absorption is less than this value at the excitation wavelength. Even with this precaution taken, it is usually a good idea to correct the fluorescence spectrum for sample absorption. Lakowicz (1983) offers the following formula for computing the corrected fluorescence of a sample at a given wavelength:

$$F_{CORR} = F_{OBS} \times 10^{(A_{ex} + A_{em}/2)}$$

where F_{CORR} is the corrected fluorescence intensity, F_{OBS} is the observed fluorescence intensity, and A_{ex} and A_{em} are the sample absorption values at the excitation and emission wavelengths, respectively.

Aside from inner filter effects, one needs to avoid too high a sample concentration also because of the limited linear response range of fluorescence detection devices. If the sample fluorescence is too great, the detector will not respond properly, and one can observe saturation effects in the spectrum. The linear dynamic range of fluorimeters varies from instrument to instrument. The instrument manufacturer should be consulted to determine the appropriate range for a particular instrument.

Collisional Quenching of Fluorescence

As described earlier, molecules emit light in an attempt to relax from an excited electronic state back to their ground state. The intensity of observed fluorescence is related to the probability of such a relaxation mechanism, which in turn depends on the alternative relaxation pathways that are available to the excited molecule. As we have seen, heat dissipation is another common means of relaxing from the excited state. Any event that increases the likelihood of such heat transfer will decrease the probability of photon emission, and thus will lower the observed fluorescence intensity. Heat dissipation can be enhanced by collisions with other molecules, especially molecules or ions with high mass. Thus, if one were to add such molecules to a sample of a fluorescent compound, the observed fluorescence would decrease with increasing concentration of the added quencher. The relationship between the observed fluorescence intensity of a sample and the concentration of quencher added is given by the Stern-Volmer relationship:

$$F_0/F = 1 + K[Q]$$

where F is the fluorescence intensity at a particular concentration of quencher, F_0 is the fluorescence intensity of the sample in the absence of any quencher, $[Q]$ is the molar concentration of quencher added to the sample, and K is a constant of proportionality known as the Stern-Volmer constant. A plot of F_0/F as a function of $[Q]$ yields a straight line with y-intercept equal to 1, and a slope equal to K. The value of K is a measure of the effectiveness of quenching and depends on the particular fluorophore, particular quencher, *and* the accessibility of the fluorophore to the quencher (Lackowicz, 1983). Figure 9.16 illustrates the results of such a quenching experiment for indole being quenched by CsCl.

The most commonly used quenchers for protein fluorescence are potassium iodide (KI) and cesium chloride (CsCl). These are typically used over a concentration range of 0 to 2 M for protein samples. KI has the disadvantage of being photolabile, and thus must be prepared immediately before use. CsCl solutions, on the other hand, are quite stable and can be prepared well in advance of the experiment. Two procedures are commonly used to carry out a quenching experiment. In the first, one sets up a series of samples of equal protein concentration and equal final

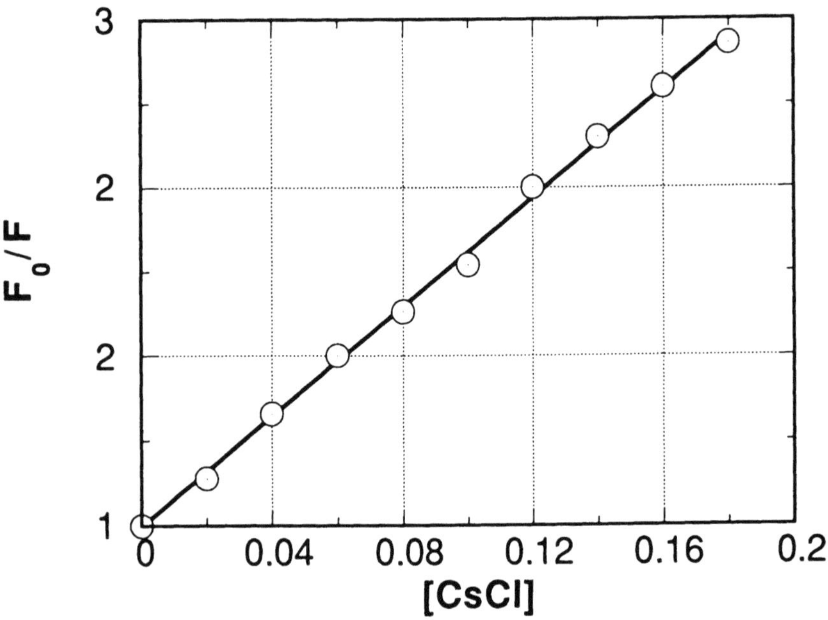

Figure 9.16 Stern-Volmer quenching plot for the quenching of indole in aqueous solution by CsCl.

volume, but with varying concentration of quencher. To perform the experiment in this fashion requires that one have a concentrated stock of protein, in sufficient quantity to prepare 5 to 10 samples. If protein supply is limiting, an alternative procedure is often used. Here one prepares a single sample of protein without quencher, and then sequentially makes additions to this sample from a concentrated stock of quencher. When using this approach one must remember to correct the measured fluorescence intensities for the dilution effects that accompany quencher addition. So, for example, if one had a 1.0 ml sample of protein to which one added 0.5 ml of stock quencher solution, one would multiply the observed fluorescence intensity by (1.0 + 0.5)/1.0 = 1.5 to get a fluorescence intensity value that is directly comparable to that of the original sample.

The value of quenching experiments to the protein scientist is that these data allow one to assess the relative solvent exposure of different types of fluorophores. The more exposed a fluorophore is, the more effective a collisional quencher will be in reducing the fluorescence intensity displayed by that molecule. In globular proteins one can often have a distribution of tryptophan or tyrosine residues between the buried interior and solvent accessible surface of the protein. Presumably the surface exposed residues would be quenched by addition of KI or CsCl, while the fluorescence of the buried residues should be unaffected by quencher addition. Consider, for example, the quenching of tryptophan fluorescence in the protein lysozyme by potassium iodide. The top panel in figure 9.17 shows the traditional Stern-Volmer plots for quenching of the native protein, and for the protein in 6 M guanidine hydrochloride. For the denatured protein, where all of the tryptophan residues are exposed to solvent, the data are well described by the linear Stern-Volmer equation. For the native protein, however, one observes a significant deviation from linearity, owing to the distribution of fluorophores within the protein interior and exposed surface for this protein. Lehrer (1971) has developed a method for determining the mole fraction of accessible fluorophores in proteins. According to Lehrer (1971), before addition of quencher, the overall fluorescence of a protein is the sum of the fluorescence from the buried and exposed populations of fluorophore:

$$F_0 = F_{0^A} + F_{0^B}$$

where F_0 is the overall sample fluorescence in the absence of quencher, and F_{0^A} and F_{0^B} are the fluorescence intensities for the accessible and buried populations of fluorophore in the absence of quencher, respectively. When quencher is added to the sample, the fluorescence of the

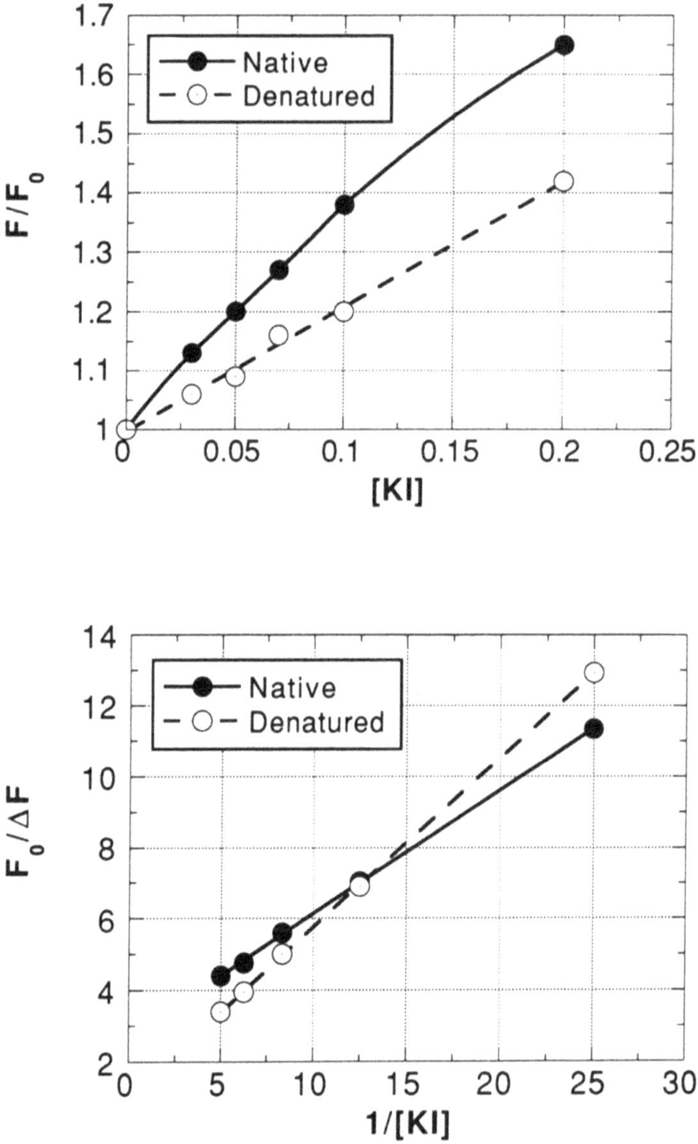

Figure 9.17 Stern-Volmer (top) and modified Stern-Volmer (bottom) plots for quenching of native and denatured lysozyme by potassium iodide.

accessible population is affected, but that of the buried population is not. Now the overall fluorescence is given by:

$$F = (F_0^A/1 + K[Q]) + F_0^B$$

Subtracting this equation from our equation in the absence of quencher we obtain:

$$\Delta F = F_0 - F = F_0^A K[Q]/1 + K[Q]$$

Noting that the mole fraction of accessible fluorophore f_A is equivalent to $F_0^A/(F_0^A+F_0^B)$, we can substitute this into the above equation and invert the equation to yield:

$$F_0/\Delta F = 1/(f_A K[Q]) + 1/f_A$$

Thus, a plot of $F_0/\Delta F$ as a function of $1/[Q]$ yields a straight line with y-intercept equal to $1/f_A$. Such a plot provides a simple method for determining the mole fraction of exposed fluorophore in a sample. The bottom panel of figure 9.17 illustrates this type of modified Stern-Volmer plot for native and denatured lysozyme. For both forms of the protein the data are well described by a linear function. The y-intercepts from these data yield estimates of the mole fraction of exposed tryptophans for native and denatured lysozyme of 0.38 and 1.00, respectively.

CIRCULAR DICHROIC SPECTROSCOPY

We saw earlier in this chapter how light can propagate as a helix with either a clockwise or counterclockwise sense of rotation. Such light is referred to as right and left circularly polarized, respectively. When two such beams of equal amplitude but opposite rotation impinge on a sample, their resultant is a plane polarized beam if the light absorption properties of the sample are equal for the two beams. If, however, the sample preferentially absorbs one rotational form over the other, the beam emerges from the sample as elliptically polarized light. The degree of ellipticity here is proportional to the difference in extinction coefficient between the left and right circularly polarized light, which in turn depends on the chirality or optical activity of the sample (Campbell and Dwek, 1984).

For proteins, the greatest use of circular dichroic (CD) spectroscopy is for assessing the distribution of secondary structure within a polypeptide backbone. In the far UV (250 to 190 nm) the amide bonds of peptides give rise to π–π^* and n–π^* electronic transitions that show optical activity. It turns out that the three major secondary structures associated with proteins—alpha helix, beta sheet, and random coil—give rise to very distinct spectral signatures in the deep UV CD. This was first demonstrated by Doty and coworkers (Holzwarth and Doty, 1965) using homopolypeptides such as poly-L-lysine. Figure 9.18 illustrates the characteristic far UV CD spectra for poly-L-lysine in its three secondary structure forms. These same spectral signatures apply to these secondary structures in globular proteins. Thus, proteins like myoglobin which are mainly composed of alpha helices display CD spectra showing the double negative features at ca. 208 and 222 nm (figure 9.19) as seen for alpha helical poly-L-lysine in figure 9.18. Proteins with high beta sheet content display a single negative feature at ca. 215 nm in their CD spectra, again much like the corresponding secondary structure form of poly-L-lysine. For most globular proteins, all three major secondary structure forms are

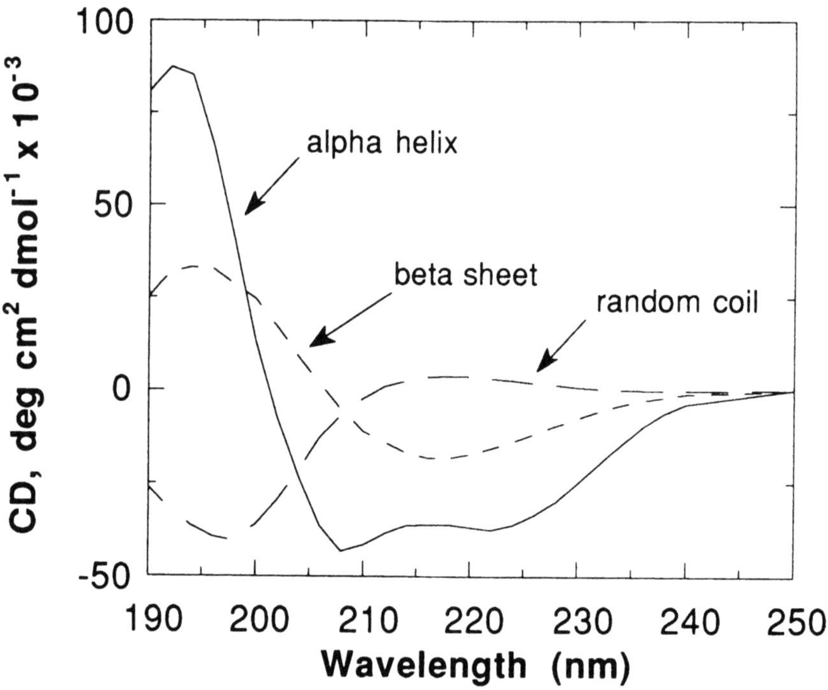

Figure 9.18 Circular dichroic spectra for poly-L-lysine in its three secondary structures.

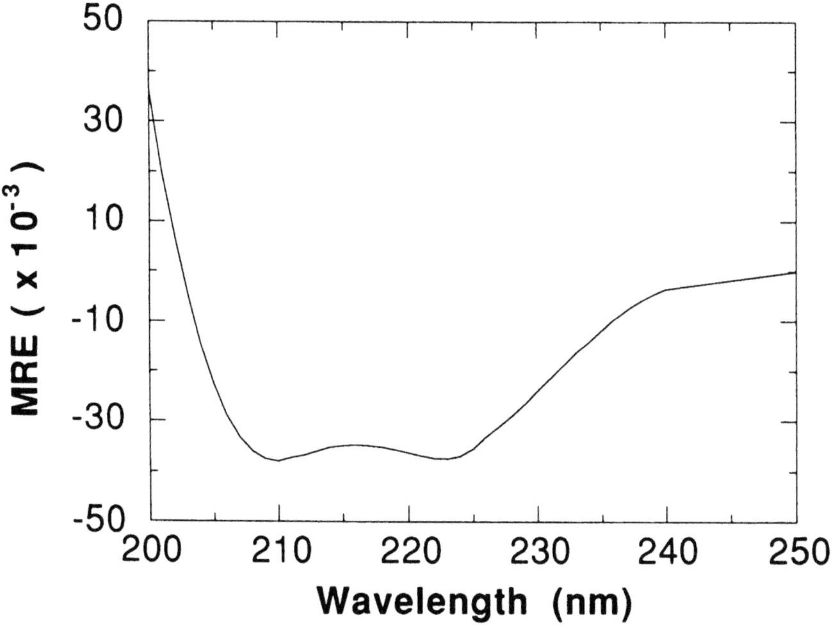

Figure 9.19 Circular dichroic spectrum for horse heart myoglobin showing a predominance of alpha helical secondary structure.

present, and their CD spectra appear as a weighted combination of the three spectra shown in figure 9.18. Methods for deconvoluting the spectrum of a globular protein into its component spectra (and hence estimating its secondary structure content) are discussed below.

Because protein CD spectra are usually collected in the deep UV, where many buffer components also absorb light, certain precautions must be taken to ensure the quality of the spectral data one obtains. Cells for CD spectroscopy must be made of high quality quartz with high light transmission properties down to 190 nm. The spectrometer and sample chamber should be purged with nitrogen gas to remove oxygen which can absorb light in the deep UV. Typically purging for 30 minutes with 5 to 15 liters/minute of nitrogen is sufficient to remove most oxygen. Of course, as with all spectroscopic measurements, the sample should be filtered to remove particulates prior to loading into the spectroscopic cell. Beyond these precautions, one must consider the non-protein components of one's sample, and their light absorption properties. Buffers vary widely in their deep UV absorption. Phosphate buffer, for example, can interfere with the acquisition of CD data below 210 nm and should thus be avoided. Tris (tris(hydroxymethyl)aminomethane) and TES (2−{[*tris*−(hydroxymethyl)methyl]amino}ethanesulfonic acid) show the least interferences in CD spectra. TES is particularly useful in this regard since it has high buffering capacity in the physiological pH range ($pK_a = 7.5$, useful pH range = 6.2–8.2). In my laboratory we have found that 10 mM TES provides adequate buffering capacity for most protein solutions, while minimizing any buffer contribution to the deep UV CD spectrum. One must also be concerned with the absorption properties of certain ions used for pH and ionic strength adjustment. Chloride ions can form charge transfer species with water and give rise to UV absorption. Thus, chloride should be avoided in samples for CD work. We have found that perchlorate and sulfate provide the best alternatives for ionic strength adjustment. Thus, a typical buffer system that one might use for CD spectroscopy of proteins might be 10 mM TES, 50 mM Na_2SO_4, pH 7.5.

Protein concentration and cell pathlength are also issues that must be considered in preparing a sample for CD work. As with absorption measurements, there is a finite range of sample concentrations over which the CD signal will respond linearly. This range varies, of course, with the pathlength of the cell by the same Beer's law relationship as for absorption spectroscopy. Often in CD work it is better to use a more concentrated protein sample and smaller pathlength cell to minimize contributions from non-protein components of the solution. Pathlengths between 0.1 and 1.0 mm are commonly used for protein CD spectroscopy. Optimal signal is obtained for protein samples with absorption values at ca. 193 nm of about 1.0. The average peptide bond in a protein has an extinction coefficient of about 7,000 M^{-1} cm^{-1} (the value is lower, about 4,000, for alpha helices). Using this value and knowing the number of residues in one's protein, one can compute a reasonable concentration for data acquisition (Yang et al., 1986). For example, if one had a protein of molecular weight 48,000 Da, and one used an average molecular weight per amino acid residue of 115, then one would calculate that the extinction coefficient for this protein at 193 nm is approximately $(48{,}000/115) \times 7{,}000 = 2.92 \times 10^6$. Taking the reciprocal of this number one would determine that a good starting concentration of protein for CD studies would be 0.3 μM for a 1 cm cell or 3 μM for a 1 mm cell. Of course these calculations are crude, but they offer the scientist a ballpark value for starting one's studies.

For the purpose of estimating secondary structure or for general comparison with literature spectra, protein CD spectra are usually reported in terms of the molar ellipticity or the mean residue ellipticity as a function of wavelength. The mean residue ellipticity is a measure of the average CD signal per residue of the protein. Most modern CD software programs (supplied by the instrument manufacturers) will convert measured CD values into molar or mean residue ellipticities if the user supplies information on the concentration of protein present, the mean residue weight for the protein (115 is used if the exact molecular weight and number of residues for the protein are unknown), and the pathlength of the cell. These and subsequent calculations depend critically on an accurate estimate of the sample concentration, so it is imperative that the user determine the concentration of protein in the CD cell. Usually colorimetric assays are too crude for this purpose. The tyrosine difference spectral method described in Chapter 3 works well for this purpose.

The accuracy of the CD data also depends on the wavelength accuracy and photometric response of the spectrometer. These issues should be addressed by the manufacturer when the instrument is purchased. However, over time, lamp intensities and other instrumental factors can change, so it is a good practice to periodically test the accuracy of one's instrument. Yang et al. (1986) recommend the use of d-10-camphorsulfonic acid in water for this purpose. This same compound can also be used to accurately determine the pathlength of one's CD cells. Details of the calibration procedure can be found in the review by these authors.

Estimating Protein Secondary Structure from CD Spectra

A number of methods for estimating the secondary structure content of proteins from their CD spectra have been described in the literature (see Yang et al., 1986 for a review). All of these methods have in common the use of a reference set of spectra that represent the spectra of each "pure" secondary structure type. For instance, some methods use the spectra of poly-L-lysine under different temperature and pH conditions, representing 100% alpha helix, 100% beta sheet, and 100% random coil as a basis set. The methods vary in the makeup of the reference set itself, and the wavelength range over which the calculations are performed. For all of these methods, one assumes that the measured CD signal for one's protein at any given wavelength λ can be expressed as the sum of the CD signals at that wavelength for the pure secondary structure types multiplied by their respective mole fractions in the proteins. If we consider a protein to be made up of alpha helices, beta sheets, and random structure only, the CD signal for the protein at any wavelength is given by:

$$\Theta_\lambda = f^\alpha \Theta_\lambda{}^\alpha + f^\beta \Theta_\lambda{}^\beta + f^R \Theta_\lambda{}^R$$

where f^i is the mole fraction of secondary structure i in the protein and $\Theta_\lambda{}^i$ is the CD signal for a sample of 100% secondary structure i at wavelength λ.

The simplest analysis of this type takes into consideration data at a single wavelength only. Consider the data presented in figure 9.18. At 208 nm the CD intensities for random coil and beta sheet structure are equal to one another, but significantly greater than that of an alpha helix. Using poly-L-lysine spectra as our reference set we find that the molar ellipticities for these secondary structures are −4,000 for beta sheet and random coil, and −33,000 for alpha helix. If we restrict our attention to only these three secondary structure types, we can estimate the mole fraction of alpha helix in a protein as follows. We assume that

$$f^\alpha + f^\beta + f^R = 1.0$$

(i.e., only these three structures contribute to the total protein). If we define

$$X = f^\beta + f^R$$

then $X = 1 - f^\alpha$. Recalling that $\Theta_{208}{}^\beta = \Theta_{208}{}^R = -4{,}000$, and that $\Theta_{208}{}^\alpha = -33{,}000$, we obtain:

$$\Theta_{208} = f^\alpha(-33{,}000) + (1-f^\alpha)\ (-4{,}000)$$

rearranging this we get:

$$f^\alpha = (\Theta_{208} + 4{,}000)/-29{,}000$$

Thus, with a single wavelength measurement we can obtain an estimate of the mole fraction of alpha helix in a protein.

The above procedure provides only crude estimates for a single secondary structure type. To increase the predictive power of our estimates and to include other secondary structure types, we must analyze the protein spectrum at multiple wavelengths. Yang and coworkers have developed computational methods for fitting the entire spectrum of a protein to a weighted sum of the three major secondary structure spectra based on a poly-L-lysine reference set. While this method provides good estimates of secondary structure, it ignores the contributions of other chiral components of proteins, such as the aromatic amino acids, that contribute to the far UV CD spectrum. For this reason several groups have developed reference sets based on the CD spectra of proteins whose secondary structure is well known from crystallographic data. Chen and Yang (1971), for instance, developed a reference set based on five proteins, while Chang et al. (1978) used a total of 15 proteins to develop their reference spectra. The use of a protein based reference set provides slightly better estimates of secondary structure for globular proteins, and has the added advantage of allowing one to estimate beta turn content as well as the three major secondary structure types. These various methods are described in detail in the recent review by Yang et al. (1986). In this review the authors provide the source code for computer programs to estimate protein secondary structure from CD spectra by several of the literature methods. Additionally, the major manufacturers of CD instruments offer software, based on one or more of these methods, for calculating protein secondary structure from CD measurements.

OTHER SPECTROSCOPIC METHODS FOR PROTEINS

In this chapter we have described a few optical methods that the generalist can use for assessing protein tertiary and secondary structure. This chapter is by no means a comprehensive review of such methods. For protein UV absorption spectroscopy, for example, second derivative methods are now commonly used to resolve the individual contributions of tyrosine and tryptophan to the absorption band envelope (Ragone et al., 1984). Such enhanced resolution allows one to independently assess structural changes in each amino acid type simultaneously. We considered only steady state measurement in discussing protein fluorescence. Fluorescence lifetime and polarization measurements, however, provide a wealth of information on protein structure and structural transitions (Lackowicz, 1983). While CD spectroscopy is the most common and familiar means of assessing protein secondary structure in solution, vibrational methods, such as infrared and Raman spectroscopy, offer excellent estimates of secondary structure without some of the interferences from buffer components that plague CD measurements (Parker, 1983). These and other spectroscopic methods are available for the study of protein structure, but are more specialized in terms of either the instrumentation or data analysis or both. Reviews of some of these methods are listed in the reference section of this chapter. Several texts have also appeared recently that are devoted exclusively to biological spectroscopy. The reader is referred to these as well for additional information on other spectroscopic methods that can be applied to proteins.

References

Campbell, I. D., and Dwek, R. A. (1984) *Biological Spectroscopy*, Benjamin/Cummings. Menlo Park, CA.

Cantor, C. R., and Schimmel, P. R. (1980) *Biophysical Chemistry, Part II*, W. H. Freeman, San Francisco, CA.

Chang, C. T.; Wu, C.-S. C.; and Yang, J. T. (1978) *Analyt. Biochem.*, **91,** 12.

Chen, R. F., and Edelhoch, H. (1975) *Biochemical Fluorescence*, Vols. 1 and 2, Marcel Dekker, New York.

Chen, Y. H., and Yang, J. T. (1971) *Biochem. Biophys. Res. Commun.*, **44,** 1285.

Copeland, R. A.; Ji, H.; Halfpenny, A. J.; Williams, R. W.; Thompson, K. C.; Herber, W. K.; Thomas, K. A.; Brunner, M. W.; Sitrin, R. D.; Yamazaki, S.; and Middaugh, C. R. (1991) *Arch. Biochem. Biophys.*, **289,** 53–61.

Demchenko, A. P. (1986) *Ultraviolet Spectroscopy of Proteins*, Springer-Verlag, New York.

Donovan, J. W. (1973) *Meth. Enzymol.*, **27,** 497–525.

Fodor, S. P. A.; Copeland, R. A.; Grygon, C. A.; and Spiro, T. G. (1989) *J. Am. Chem. Soc. USA*, **111,** 5509.

Freifelder, D. (1982) *Physical Biochemistry*, W. H. Freeman, San Francisco, CA.

Herskovits, T. T., and Sorensen, M. (1968) *Biochemistry*, **7,** 2533–2542.

Holzwarth, G. N., and Doty, P. (1965) *J. Am. Chem. Soc.*, **87,** 218.

Kotake, S.; Hey, P.; Mirmira, R. G.; and Copeland, R. A. (1991) *Arch. Biochem. Biophys.*, **285,** 126–133.

Lackowicz, J. R. (1983) *Principles of Fluorescence Spectroscopy*, Plenum Press, New York.

Lehrer, S. S. (1971) *Biochemistry*, **10,** 3254–3263.

Parker, F. S. (1983) *Applications of Infrared, Raman, and Resonance Raman Spectroscopy in Biochemistry*, Plenum Press, New York.

Ragone, R.; Colonna, G.; Balestrieri, C.; Servillo, L.; and Irace, G. (1984) *Biochemistry*, **23,** 1871–1875.

Yang, J. T.; Wu, C.-S. C.; and Martinez, H. M. (1986) *Meth. Enzymol.*, **130,** 208–269.

Zabin, H. B. (1991) Ph.D. Thesis, The University of Chicago, Chicago, Illinois.

Zabin, H. B., and Terwilliger, T. C. (1991) *J. Mol. Biol.*, **219,** 257–275.

10

Protein Folding and Stability

We have seen in Chapter 1 that the biological activity of a protein is directly dependent on the adoption of a specific three-dimensional structure by the polypeptide chain. This correctly folded structure, referred to as the native conformation, is thought to represent not only the most biologically active form of the protein, but also an arrangement of the polypeptide chain of minimal potential energy. While it is true that there is some thermodynamic stability to the native conformation of a protein relative to the unfolded state, this stabilization can be quite modest. It is surprising, but true, that for many proteins the native conformation is stabilized over the denatured form by only a few kcal/mol. Thus it is easy to see why protein stability is often a problem for long term storage or use of proteins in solution. In Chapter 2 we discussed general methods for maximizing the stability of proteins for laboratory purposes. In this chapter we shall explore the use of studying protein unfolding by chemical and thermal means as a measure of the relative stabilities of different proteins or the same protein under different solution conditions.

PROTEIN DENATURATION AND THE TWO STATE MODEL

Over the past 40 years or so, a rich literature on protein unfolding (denaturation) and refolding has accumulated. In these studies scientist have probed the pathways by which proteins go from their compact native conformations to the fully unfolded state that is attained in high concentrations of chemical denaturants. For the vast majority of single subunit proteins the results of these studies suggest that the unfolding event can be viewed as an equilibrium process involving two states, the fully native and the fully denatured states of the protein (Pace, 1975):

$$N \rightleftarrows D$$

where N and D represent the native and denatured states of the protein respectively. Thus under equilibrium conditions we need only consider these two states of the protein. Any intermediate states that might be accessed kinetically do not accumulate to a sufficient population that they need to be considered for our purposes. If this assumption holds, then the total population of protein molecules in solution is the sum of the mole fractions of native molecules and denatured molecules, i.e.:

$$1.0 = f_D + f_N$$

or

$$f_N = 1.0 - f_D$$

and

$$f_D = 1.0 - f_N$$

These equalities allow us to define an equilibrium constant, and hence a Gibb's free energy for the unfolding or folding reaction in terms of these mole fractions:

$$K = [D]/[N] = f_D/(1 - f_D)$$

and

$$\Delta G = -RT \log(K)$$

where R and T are the ideal gas constant and the temperature in Kelvins, respectively. Thus if the two state model applies to a particular protein we can define the tendency for the protein to denature (i.e., the inherent stability of the native protein) in familiar thermodynamic terms.

How well does this two state model hold for protein denaturation in general? As stated above it describes quite accurately the denaturation behavior for most single subunit proteins. The model often fails, however, for multisubunit proteins and for integral membrane proteins, where the membrane spanning sections tend to denature independently of the aqueous exposed domains. The two state model is also often inadequate for describing the thermal denaturation of proteins because of interfering processes that take place at elevated temperatures; these will be discussed below.

APPLICATION OF THE TWO STATE MODEL TO CHEMICAL DENATURATION OF PROTEINS

Several chemical agents are known to induce the unfolding of native proteins when they are present at high concentrations. Among these, urea and guanidine hydrochloride are the most commonly used chemical denaturants. Titration of a protein solution with one of these chemicals shifts the equilibrium between the native and denatured states of the protein to gradually favor the denatured form. The amount of denaturant needed and the steepness of the titration curve provide us with quantitative information on the stability of the protein sample. Let us work through an illustrative example of how one might perform and analyze a protein denaturation experiment. In this example we shall follow the guanidine hydrochloride induced unfolding of bovine serum albumin by measuring the UV difference spectrum of the protein solutions at varying guanidine concentration.

MATERIALS

1. 10 mg/ml BSA solution (BSA from Sigma, Bio-Rad, Pierce, etc.)
2. 6 M Guanidine hydrochloride (use low UV absorbing grade material available from Pierce)
3. Buffer.

PROCEDURE

1. To each of 12 plastic test tubes add the portions of BSA, buffer, and 6 M guanidine hydrochloride (Gdn) shown in table 10.1 to reach the indicated final concentrations of Gdn.
2. After running a spectrum of a buffer blank, record and store the UV spectrum of the contents of each tube from 350 to 250 nm as described in Chapter 9. Compute the difference spectrum by subtracting the spectrum of tube 1 (0 M Gdn) from the spectrum of each other tube. The wavelength maximum for the difference spectra will occur at ca. 287 nm. Plot the ΔA value at this wavelength as a function of Gdn concentration.

Table 10.1 Experimental set up for performing a guanidine titration of BSA.

Tube No.	[Gdn], M	μl of 10 mg/ml BSA	μl of 6 M Gdn	μl of Buffer
1	0	100	0	1,000
2	0.5	100	83	817
3	1.0	100	167	733
4	1.5	100	250	650
5	2.0	100	333	567
6	2.5	100	417	483
7	3.0	100	500	400
8	3.5	100	583	317
9	4.0	100	667	233
10	4.5	100	750	150
11	5.0	100	833	67
12	5.4	100	900	0

Data Analysis

Figure 10.1 illustrates the results of a typical titration of BSA with Gdn. The solid line drawn through the data represent a non-linear least squares fit of the data to the two state model described above. The maximum value of ΔA occurs above ca. 4 M Gdn, and has an average value (based on the data points in this plateau region) of 0.11138. Now for any experimental measurement X (including CD, fluorescence, UV absorption, etc.) if we can define a unique value of that experimental measure for the denatured state of the protein X_D and another unique value of that measurement for the native state of the protein X_N, then the value of that measurement for any particular solution condition will be given by:

$$X = f_D X_D + f_N X_N$$

or:

$$X = f_D X_D + (1 - f_D) X_N$$

We can rearrange this to give an equation for the mole fraction of denatured protein under any solution conditions:

$$f_D = X - X_N / X_D - X_N$$

Now in the present example our experimental measurement is $\Delta A_{287\ \text{nm}}$, and the values for X_N and X_D are zero and 0.11138, respectively. Thus, for this example:

$$f_D = \Delta A/0.11138$$

Using this equation we can re-graph our data as shown in figure 10.2, where the mole fraction of denatured protein goes from a minimum of zero at low Gdn concentration to a maximum of 1.0 above a critical Gdn concentration. From a curve like this one can determine the concentration of Gdn required to reach 50% denaturation, i.e., where $f_D = 0.50$; this value is referred to as c_m (the midpoint concentration). For the data in figure 10.2 this occurs at about 2.6 M Gdn. This concentration required for half denaturation should be an intrinsic property of the protein under the specific temperature, pH, and other solution conditions used in the experiment. Multiple samples of the same protein should, in principle,

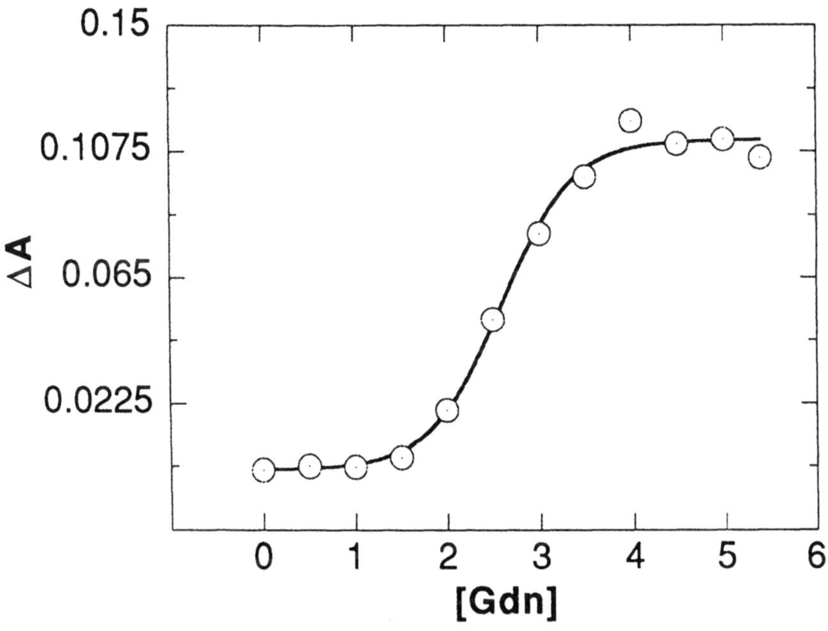

Figure 10.1 Plot of absorption difference at 287 nm (ΔA) for bovine serum albumin (BSA) as a function of guanidine hydrochloride (Gdn) concentration. ΔA was determined from the difference spectrum of the samples at various Gdn concentrations minus that of a sample without any denaturant.

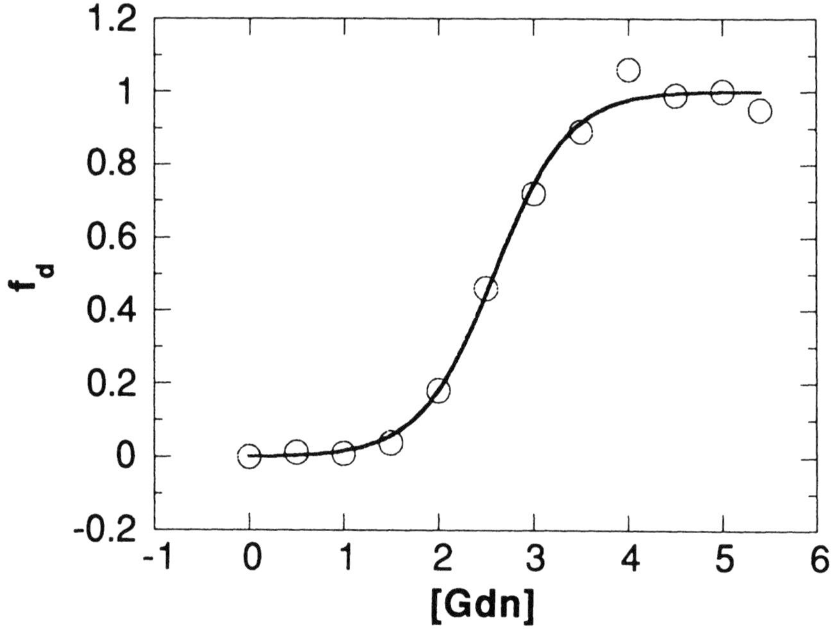

Figure 10.2 Plot of fractional denaturation (f_D) for BSA as a function of Gdn concentration for the data presented in figure 10.1.

yield the same denaturation curves, assuming that the samples all start out with the protein in its native state. If, on the other hand, a particular sample has a higher proportion of denatured protein at the start, the apparent midpoint of the Gdn or urea titration will occur at a lower denaturant concentration. Thus, denaturant titration curves provide us with a means of assessing the relative level of protein stability for different samples. These can be used, for example, to compare the relative stability of sample of a particular protein under different solution conditions, or to compare the relative stabilities of different proteins. The effects of point mutations on overall protein stability have been studied in this fashion for a number of proteins recently. For example, Terwilliger and coworkers have used such denaturation studies to assess the effects of point mutations on the stability of the bacteriophage gene V protein (Liang and Terwilliger, 1991), Kellis et al. (1988) have performed similar studies for the enzyme barnase, and several groups have studied point mutations of the enzyme lysozyme. These are just a few examples of the many applications of denaturation studies that have appeared in the recent literature.

Recall that we can define an equilibrium constant for the denaturation reaction in terms of the mole fraction of denatured protein present (see above). Thus, from the data in the transition region of figure 10.2 one can calculate an equilibrium constant, and hence a Gibbs free energy for the denaturation reaction as a function of denaturant concentration. Figure 10.3 plots the apparent free energy as a function of Gdn concentration for the unfolding of BSA using the data from the transition region of figure 10.2. Note that the apparent ΔG values depend linearly on the concentration of denaturant over the range of concentrations considered in figure 10.3. Pace (1975) and others have suggested that this linear relationship can be extrapolated back to zero molar denaturant to obtain an estimate of the intrinsic stability of the protein. In other words, the

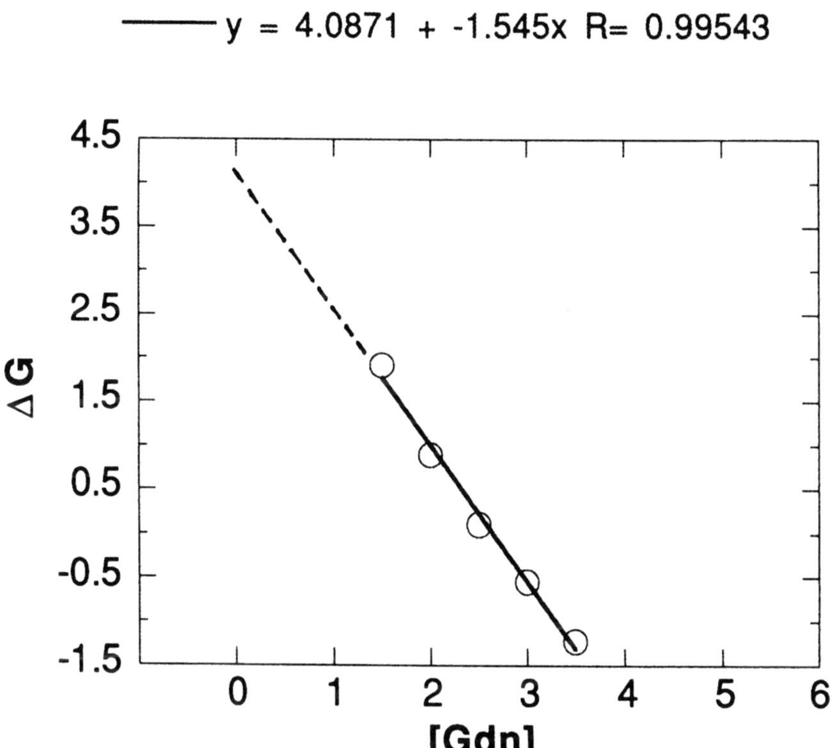

Figure 10.3 Plot of the free energy of unfolding (ΔG) as a function of Gdn concentration for the data from figure 10.1. The solid line through the data represents the linear least squares best fit to the data. The dashed line extends this linear fit to zero denaturant concentration.

extrapolated value of ΔG at zero molar denaturant, ΔG_{H_2O}, represents the thermodynamic spontaneity of the protein unfolding in the absence of denaturant. How reasonable is it to assume that the linear behavior seen in figure 10.3 extends over a wide concentration range down to zero molar denaturant? For most proteins this linear behavior does not in fact hold over any extended range of denaturant concentrations (Pace, 1975). Nevertheless, the extrapolated values of ΔG_{H_2O}, are widely used to gauge the relative stabilities of different proteins or of a protein under different solution conditions. Of course one need not extrapolate to zero molar denaturant to compare free energies of unfolding for different proteins. Recently, several groups have begun to report ΔG_x values, where X is some arbitrary denaturant concentration than is within the measurable linear denaturant concentration range for the protein. The advantage of such values is that one eliminates the implicate assumptions used in the extrapolation method.

Pace (1975) suggests that one should report two values to define the stability of a protein; the denaturant concentration required to reach half maximal unfolding, c_m, and the ΔG_x value. Together these two values uniquely define the denaturation behavior of the protein, and allow for accurate comparisons of stabilities among proteins.

THERMAL DENATURATION OF PROTEINS

Anyone who has ever fried an egg knows that proteins are susceptible to denaturation at elevated temperatures. In principle, thermal denaturation should lead to the same fully unfolded state that one attains at high concentrations of chemical denaturants. Often, however, one finds that the physico-chemical properties of the thermally denatured protein differ from those of the chemically denatured protein. For example, figure 10.4 compares the fluorescence spectra of bovine arrestin, a protein of the visual signal transduction cascade, in its native, Gdn denatured, and thermally denatured forms (Kotake et al., 1991). Note the difference in spectral properties between these three forms of arrestin.

Another difference between thermal and chemical denaturation of proteins is that, unlike chemical denaturation, thermal denaturation is often irreversible (as in our example of the fried egg). The reason for this is that the simple two state equilibrium that held for chemical denaturation often is not sufficient to describe the process of thermal denaturation. As the temperature of a protein solution increases, the equilibrium between the folded and unfolded states shifts in favor of the unfolded

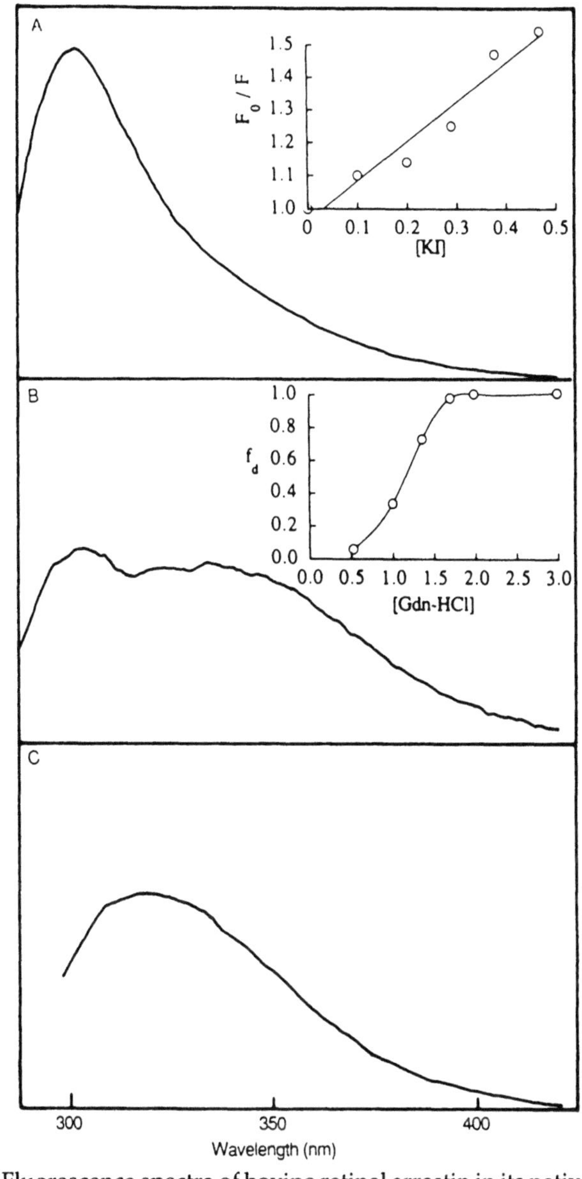

Figure 10.4 Fluorescence spectra of bovine retinal arrestin in its native conformation (A), after treatment with 4 M Gdn (B), and after thermal denaturation by incubation at 79° C for five minutes. The insets to panels A and B show the Stern-Volmer quenching of tyrosine fluorescence by KI, and the effects of titrating the protein with Gdn, respectively. From Kotake et al. (1991), with permission.

form. In this state, hydrophobic residues that are normally shielded from the polar solvent in the folded molecule become exposed to solvent. The thermodynamic cost of the repulsive interactions between these residues and the solvent are significant enough to drive the spontaneous refolding of proteins when solution conditions are adjusted to those favoring the folded protein. For example, when unfolded proteins in 6 M Gdn are diluted with buffer to bring the Gdn concentration down, one often finds that the proteins refold into their proper native structure spontaneously. Elevated temperatures, however, not only favor the unfolded state of a protein, but also greatly increase the rate of collisions between unfolded protein molecules. These collisional encounters between unfolded monomeric proteins provide an alternative means of occluding hydrophobic residues from solvent by formation of protein oligomers. In extreme cases, this oligomerization continues until large protein aggregates form which then fall out of solution as precipitates. Once formed, these oligomeric protein particles are slow to repartition into unfolded monomers. Thus, at high temperatures the equilibrium between the folded and unfolded monomeric protein is greatly perturbed by the competing process of oligomer formation.

Despite the irreversibility of thermal denaturation studies, they have nevertheless been widely used to study the relative thermal stability of different proteins (Privalov and Khechinashvili, 1974). To obtain reproducible results from thermal denaturation studies, it is important that one vary the temperature of a protein solution in small increments (say 5° C or smaller steps) and allow sufficient time at each temperature for the solution to reach thermal equilibrium (at least 5 minutes for a 1 to 3 ml sample). With these precautions taken, one can assess the thermal unfolding of proteins by any of a variety of methods. Spectroscopic and biological activity assays are most commonly used for this purpose. In fact, many spectrometers are now available with thermostatic sample holders that allow one to vary the sample temperature by attaching the sample block to a circulating water bath. One important point to keep in mind, however, is that temperature can affect the spectroscopic properties of molecules in ways independent of protein unfolding. For example fluorescence quantum yields tend to decrease with increasing temperature for all fluorophores, regardless of the conformation of the protein. Because of the usually irreversible nature of thermal denaturation of proteins, one can often perform these studies by incubating samples at different temperatures for a fixed period of time, and then cooling the samples to room temperature before collecting the spectroscopic or biological activity assay. As an example of this strategy my laboratory has studied the thermal denaturation of bovine arrestin by a number of criteria (Kotake et al., 1991). For these studies, a sample of the protein was divided into 25 equal aliquots and these were placed on ice. Using a variable temperature water bath, each sample was incubated for 5 min at a different temperature between 25° and 80° C. After incubation the samples were cooled to 25° C, and compared by deep UV circular dichroism (to assess secondary structure content), fluorescence spectroscopy (to assess tertiary structure), and by a receptor binding assay (to assess biological activity). In figure 10.5 the thermal denaturation curves for these three assays are compared for arrestin. There are two important points to glean from this figure. First, the thermal denaturation of arrestin

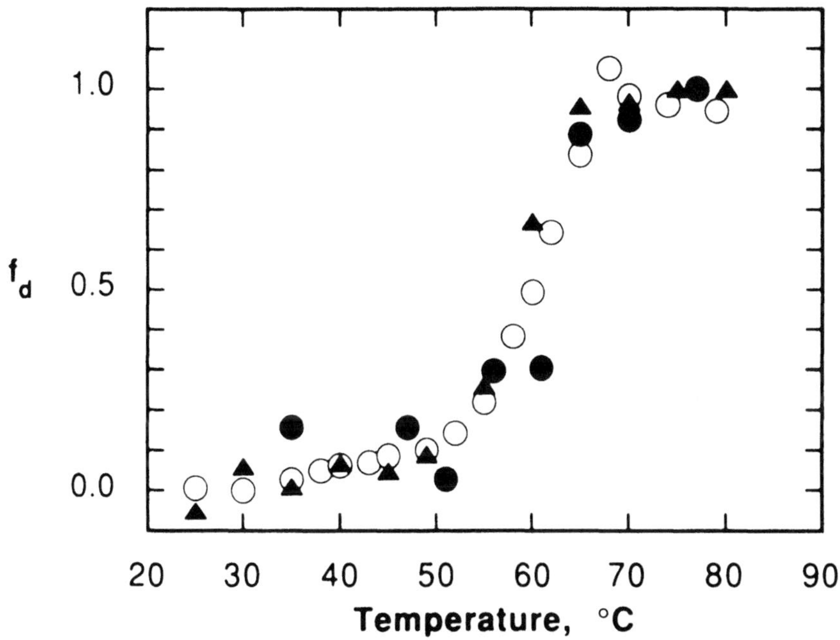

Figure 10.5 Thermal denaturation curves for bovine retinal arrestin monitored by three reporters of protein structure. The open circles represent the effects of temperature on the fluorescence spectrum of the protein, which reports changes in tertiary structure. The closed circles represent the effects of temperature on the ability of the protein to bind to its receptor, light adapted phosphorylated rhodopsin, which reports changes in biological activity. The closed triangles represent the effects of temperature on the circular dichroic intensity at 214 nm, which reports changes in secondary structure. All three data sets indicate an unfolding midpoint temperature of 59 ± 2° C for this protein. From Kotake et al. (1991) with permission.

appears to be irreversible on the time scale of these experiments, since all three data sets follow the type of two state curve that one expects. Second, all three data sets appear to overlay one another, indicating that secondary structure, tertiary structure, and biological activity are all lost simultaneously in the same unfolding reaction. The midpoint temperature, i.e., the temperature at which 50% of the protein is denatured, was determined for each data set in figure 10.5. All three data sets give similar values for the midpoint temperature, 59 ± 2° C. The excellent agreement between these independent measures of the thermal denaturation of arrestin attest to the validity of the protocol used for these studies, and further indicate that arrestin thermally denatures in a simple two state fashion.

As with chemical denaturation, thermal denaturation of proteins can be followed with any probe that discriminates between the native and denatured states. In the example above, CD, fluorescence, and receptor binding were used to follow the thermal denaturation of arrestin. Other methods, such as UV difference spectroscopy, can also be used to follow thermal denaturation of proteins. Regardless of the probe used, however, it is important that the reproducibility of the method be evaluated. One can observe widely disparate results in thermal denaturation studies, depending on the rate of temperature change used, and the amount of time that samples are allowed to equilibrate at each temperature.

Another method that is finding increased use for studying the thermal denaturation of proteins is differential scanning calorimetry (DSC; Privalov and Khechinashvili, 1974). In this method, a sample of protein solution and an identical solution lacking protein (a buffer blank) are loaded into separated ampoules or sample chambers of a highly sensitive calorimeter. The temperature of the samples is then slowly varied, and the heat capacities of the samples are measured electronically. After correcting for changes in the buffer sample, the change in heat capacity of the protein (ΔC_p) is plotted as a function of temperature. Figure 10.6 illustrates the results of a typical DSC run for the thermal denaturation of a protein. Below the transition temperature T_m, the heat capacity of the protein varies little, and is representative of the native state of the protein. Near the transition temperature one observes a large change in heat capacity as the protein thermally unfolds. After the transition, one again observes a relatively constant ΔC_p, although not necessarily the same value as before the transition; this value is representative of the heat capacity of the denatured protein. In general, one finds that denatured proteins display about 1–4 kcal/mol greater heat capacity than that of their native conformations (Pace, 1975).

Recall that ΔC_P can be related to the Gibb's free energy change for

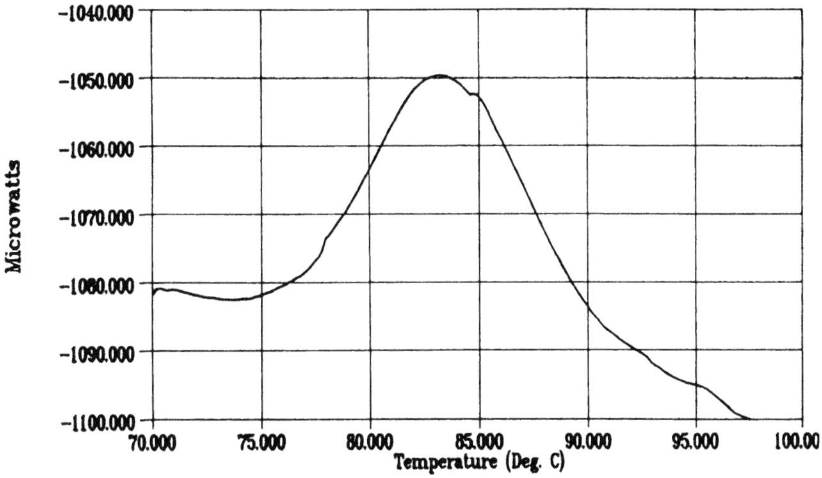

Figure 10.6 Differential scanning calorimetry trace for the thermal denaturation of horse heart cytochrome *c* at pH 4.4. Data provided by Stephen R. Lynch, Department of Biochemistry and Molecular Biology, The University of Chicago.

an equilibrium process. Thus, more familiar thermodynamic quantities, such as enthalpy (ΔH) and entropy (ΔS) changes associated with thermal denaturation can be determined from DSC data. Such information can be of great value in comparing the thermal stability of different proteins, such as point mutations of a naturally occurring protein. A number of studies of this type have appeared in the literature. DSC has been widely used to study protein thermal stability since the 1970's. At present there are two companies within the United States that offer DSC units for use with proteins. These are Hart Scientific (Utah) and Microcal (Massachusetts). More information on the use of DSC for monitoring protein stability can be obtained from either of these companies. Reviews of the application of DSC to studies of protein stability have also recently appeared; these are listed at the end of this chapter (Privalov and Khechinashvili, 1974; Sturtevant, 1987).

The problem of the irreversibility of thermal denaturation of proteins can sometimes be ameliorated by the addition of a chemical denaturant at relatively low concentrations. At concentrations below that needed to affect unfolding at ambient temperatures, chemical denaturants can lower the unfolding temperature of a protein. Since the protein unfolds at a lower temperature, the collisional processes that lead to aggregation at elevated temperatures are slowed down. In some cases this strategy has allowed researchers to perform reversible thermal denaturation on proteins that otherwise would irreversibly aggregate upon thermal denaturation. The addition of chemical denaturants also seems to help keep soluble denatured protein forms that otherwise would fall out of solution. Several groups have applied this strategy successfully to a variety of proteins (Pace, 1975).

METHODS FOR FOLLOWING PROTEIN AGGREGATION

One of the more common problems observed during long term storage of protein solutions is the formation of large aggregates of protein that fall out of solution as precipitates. The large, macroscopically observable protein particles result from the gradual build up of smaller protein aggregates that are less easily assessed by visual inspection of the sample. The progenitors for these aggregates are themselves smaller multimers of protein monomers that form either through the formation of intermolecular disulfide bonds under oxidizing conditions, or through hydrophobic-hydrophobic interactions between unfolded, or partially unfolded protein monomers. Some steps that can be taken to reduce the rate of formation of such aggregates were discussed briefly in Chapter 2. Here we shall discuss analytical methods for detecting the presence of protein aggregates at early stages in their development.

The methods most commonly used to detect aggregate formation are based on discriminating aggregated proteins from monomers on the basis of size differences or on the basis of the increased light scattering properties of larger protein aggregates. Methods based on size discrimination include native gel electrophoresis, and size exclusion chromatography; these methods have been described in Chapters 4 and 7, respectively.

Light scattering methods provide a sensitive measure of aggregate formation in protein solutions. Several commercial instruments are now available that are specifically designed for light scattering measurements. However, a common fluorimeter can be conveniently used to perform these measurements for the present purpose. Most commercial fluorimeters are designed so that the emitted light from the sample is collected at 90° from the incident excitation beam. This same design can effectively be used to measure light scattering from a sample if the excitation and emission monochromators are set to similar wavelengths. The intensity of light scattering is proportional to $1/\lambda^4$ (where λ is wavelength) so that greater sensitivity in measuring light scattering is achieved by going to lower wavelength excitation light. Typically, 350 nm light is used in these measurements because this is a wavelength at which proteins show little absorption or excitation of fluorescence. To avoid damage to the photodetector of one's fluorimeter, it is a good practice to displace the emission monochromator wavelength slightly from that of the excitation wavelength. For example, a reasonable set up would be to have the excitation monochromator set at 350 nm and the emission monochromator set at 355 nm. Using such a set up, one can then measure the relative light scattering of different protein solutions to gauge the relative amounts of aggregation in these samples.

Suppose that one wished to assess the long term stability of a protein in solution at 25° C. One might choose to measure the light scattering of such a solution at different time points over the course of several days or weeks. One difficulty with such an experiment is that the apparent light scattering intensity may vary from day to day due to changes in instrument response, rather than because of true changes in the sample. To correct for such instrument variation, it is important that one run some standard samples each day and normalize the measured light scattering of the samples to these standards. Distilled water that has been filtered through a 0.22 μ filter serves well as a low scattering standard. For a high light scattering sample one can use a suspension of corn starch or barium sulfate in water. These suspensions should be made up at a specific mass to volume ratio (typically 0.01 to 0.1 g per 100 ml) and must be stirred during the measurements to avoid particle settling. To normalize the scattering intensity of one's sample the following equation can be used:

$$I = I_S - I_{H_2O}/I_{Std} - I_{H_2O}$$

where I is the normalized light scattering intensity of ones sample, I_S is the measured intensity of ones sample, and I_{H_2O} and I_{Std} are the measured intensities for water and the high scattering standard suspension, respectively.

Alternatively, one can measure light scattering of ones sample simultaneously with collection of the sample's tryptophan fluorescence spectrum. As discussed in Chapter 9, one typically excites a protein sample at 280 nm and measures the emitted light from say 290 to 400 nm. If, however, one starts collecting emitted light closer to the excitation wavelength, say at 285 nm, one will observe strong light scattering for samples containing aggregated protein. In such an experiment, the intensity of tryptophan fluorescence from the sample can serve as a rough internal intensity standard for normalizing the light scattering intensity measurement. As an illustration of this method, consider the following experiment. Samples of a 1 mg/ml solution of bovine serum albumin were incubated in a boiling water bath for various lengths of time. After incubation the samples were allowed to equilibrate at room temperature for several hours. The samples were then excited with 280 nm light, and their emission spectra were recorded from 285 to 400 nm.

With increasing time in the boiling water bath the chances of aggregate formation increase significantly. The fluorescence spectra showed a steady increase in intensity at 285 nm with increasing incubation time. In figure 10.7 we plot the light scattering intensity at 285 nm, normalized to the fluorescence intensity for the protein at 330 nm, as a function of incubation time in boiling water. It is clear from this figure that one can use this method to easily discern the presence of aggregated protein in sample solutions.

REFOLDING OF DENATURED PROTEINS

Earlier in this chapter we noted that the unfolding of a protein can be considered as an equilibrium process between the native and denatured states of the molecule. Like any other chemical equilibrium, the unfolding reaction of a protein is reversible, at least under a particular set of condi-

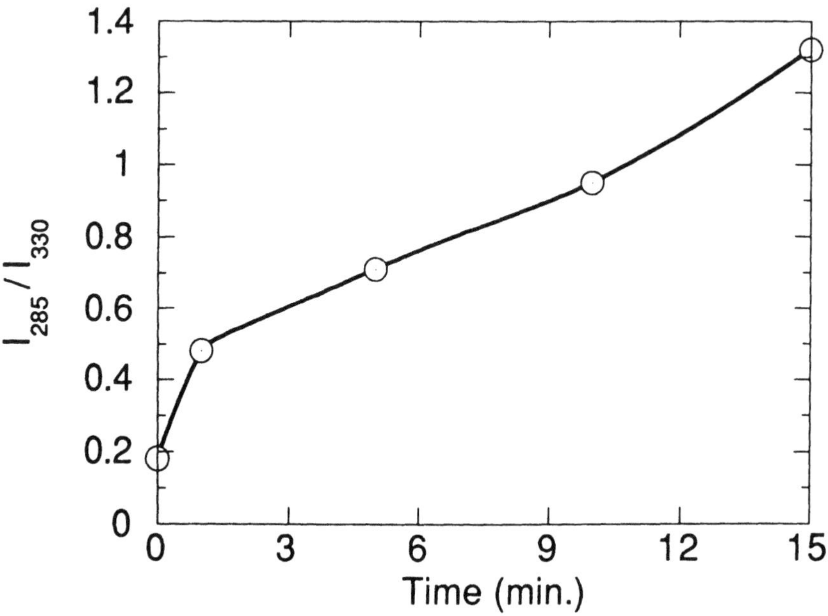

Figure 10.7 Light scattering of BSA solutions as a function of incubation time in a boiling water bath. The light scattering was measured during the acquisition of the fluorescence spectrum of the protein as described in the text. As more protein aggregation occurs in solution with longer incubation times, the light scattering intensity increases relative to the intrinsic fluorescence of the protein at 330 nm.

tions. This implies that if we somehow denature our target protein, there is a good chance that, with the proper adjustments of solution conditions, we can push the equilibrium in favor of the folded state of the protein and thus refold it. Suppose, for example, we denature a protein in 6 M Gdn and then dilute the solution with buffer to reduce the Gdn concentration to, say, 0.5 M. For most single subunit proteins, this reduction in denaturant concentration will lead to the spontaneous refolding of the protein. This phenomenon has been known for many years, and has lead to the idea that all of the information needed to define the correct three-dimensional structure of a protein is encoded within the amino acid sequence of the protein. This amazing property of proteins, and the details of how amino acid sequence defines protein folding patterns, remain an extremely interesting and active area of research. This phenomenon also leads to a very practical advantage for the protein scientist, in that it leads to a means of refolding denatured proteins to regain their active structures.

The key to successful refolding of proteins is to first establish conditions where all of the protein molecules are homogeneously in the denatured state, and then to slowly change those solution conditions to those that best stabilize the native conformation. This is typically done by dissolving the protein sample into a solution containing high concentrations of denaturant (usually urea or Gdn) and disulfide reducing agents (typically dithiothreitol). The sample is then loaded into a dialysis bag and dialyzed against a solution of lower denaturant concentration. The dialysis is sequentially repeated against decreasing concentrations of denaturant until one finally dialyzes the sample against the buffer without any denaturant. While time consuming, this method works well for a large number of proteins. The details of how one performs this refolding process will vary from protein to protein. Below is a general protocol for refolding proteins that has worked well in my laboratory. It serves as a good starting point; one may, however, need to adjust denaturant concentrations and dialysis times to optimize the refolding of one's particular target protein.

MATERIALS

1. Stock solution of 6M Gdn
2. Stock solution of 1 M DTT
3. Dialysis tubing prepared as described in Chapter 2
4. Protein sample
5. Magnetic stir plate and stir bar
6. A 1 liter beaker, flask, or other container in which to perform the dialysis.

PROCEDURE

1. Dissolve the protein to a concentration of 10 mg/ml or less in a solution of 6M Gdn, 5 mM DTT in the buffer system of choice. If the sample is an aggregated protein mass, one may need to suspend it in the Gdn solution and sonicate to facilitate dissolution. Load this into dialysis tubing, and seal the tube at both ends.
2. Place the dialysis bag in a beaker containing a stir bar and 200 volumes of a solution of 4M Gdn, 1mM DTT in one's buffer system (typically one might use a 5 ml protein sample and dialyze against a liter of solution). Place the beaker on a magnetic stir plate within a 4° C cold room or refrigerator and begin stirring slowly. Cover the beaker and allow dialysis to occur for 2 h.
3. Carefully decant the solution in the beaker and replace it with an equal volume of a buffer solution containing 2.5 M Gdn. Dialyze the protein sample as above for 2 h.
4. Carefully decant the solution in the beaker and replace it with an equal volume of a buffer solution containing 1.0 M Gdn. Dialyze the protein sample as above for 2 h.
5. Carefully decant the solution in the beaker and replace it with an equal volume of buffer solution (free of denaturant). Dialyze the protein sample as above overnight (at least 10 h).
6. Remove the sample from the dialysis bag and centrifuge or filter to remove any undissolved material. The protein should be largely refolded at this point.

This basic strategy of slowly decreasing denaturant concentration to refold proteins has been successfully applied to a very large number of proteins that have become denatured as a result of expression in foreign organisms, harsh purification steps, etc. One of the most common uses for this type of strategy is for solubilizing and refolding aggregated proteins that result from overexpression in bacterial hosts or that result from inadvertent thermal denaturation (Pace, 1975).

References

Kellis, J. T.; Nyberg, K.; Sali, D.; and Fersht, A. R. (1988) *Nature*, **333**, 784–786.

Kotake, S.; Hey, P.; Mirmira, R. G.; and Copeland, R. A. (1991) *Arch. Biochem. Biophys.*, **285**, 126–133.

Liang, H., and Terwilliger, T. C. (1991) *Biochemistry*, **30**, 2772–2782.

Pace, C. N. (1975) *CRC Crit. Rev. Biochem.*, **3**, 1–43.

Privalov, P. L., and Khechinashvili, N. N. (1974) *J. Mol. Biol.*, **86**, 665–684.

Sturtevant, J. M. (1987) *Annu. Rev. Phys. Chem.*, **38**, 463–488.

Appendix I
Glossary of Common Abbreviations in Protein Science

A–The one letter abbreviation for the amino acid alanine.

A or ABS.–Absorbance, a measure of the light absorption propensity of a molecule (see chapter 9).

a.a.–Amino acid.

Å–Angström, a measure of distance equal to 10^{-10} meters, used in spectroscopy, crystallography, and as a convenient measure of distance on the molecular scale.

Ala–Three letter abbreviation for the amino acid alanine.

Arg–Three letter abbreviation for the amino acid arginine.

Asp–Three letter abbreviation for the amino acid aspartic acid.

Asn–Three letter abbreviation for the amino acid asparagine.

BCA–Bicinchoninic acid, a colorimetric reagent used in the determination of protein concentration (see chapter 3).

Blotto–A solution of bovine serum albumin and/or non-fat dry milk used to block open sites on nitrocellulose for immunoblotting (see chapter 5).

BSA–Bovine serum albumin.

C or Cys–One and three letter abbreviations for the amino acid cysteine.

C_α–The alpha carbon that is common to all amino acid structures.

CD–Circular dichroism, a spectroscopic technique commonly used to estimate protein secondary structure (see chapter 9).

Ci–Curie, a measure of radioactivity.

CMC–Critical micellization concentration, the concentration above which any detergent molecules added to solution will all be incorporated into a micelle structure (see chapter 2).

cm^{-1}–Reciprocal centimeters or wavenumbers, a measure of frequency commonly used in electronic and vibrational spectroscopy (see chapter 9).

CNBr–Cyanogen bromide, a chemical reagent that cleaves peptide bonds selectively on the C-terminal side of methionine residues (see chapter 7).

CPM–Counts per minute, a relative measure of radioactivity or photon flux.

CZE–Capillary zone electrophoresis, an electrophoretic method for separating proteins and other biopolymers (see chapter 4).

D–One letter abbreviation for the amino acid aspartic acid.

Da–Dalton, a measure of molecular mass.

DTT–Dithiothreitol, a disulfide bond reducing agent.

E–One letter abbreviation for the amino acid glutamic acid.

$E_{cm}^{1\%}$–The absorbance of a 1% (w/v) solution of a protein contained in a 1 cm pathlength cell (see chapter 9).

ELISA–Enzyme-linked immunosorbent assay (see chapter 5).

ε–The extinction coefficient or molecular absorptivity of a molecule.

F–One letter abbreviation for the amino acid phenylalanine.

F–Fluorescence, a measure of the relative light emission from a molecule.

F_{ab}–The antigen binding domain of an antibody molecule (see chapter 5).

F_c–The domain of an antibody molecule composed of the two disulfide-linked heavy chains (see chapter 5).

f_d–Fractional denaturation, that portion of the total protein in a sample that is in the denatured or unfolded state (see chapter 10).

f_N–That portion of the total protein in a sample that is in the native or folded state (see chapter 10).

G or Gly–One and three letter abbreviations for the amino acid glycine.

Glu–Three letter abbreviation for the amino acid glutamic acid.

Gln–Three letter abbreviation for the amino acid glutamine.

H or His–One and three letter abbreviations for the amino acid histidine.

HPLC–High performance liquid chromatography (see chapter 4).

I or Ile–One and three letter abbreviations for the amino acid isoleucine.

IEC–Ion exchange chromatography (see chapter 4).

IEF–Isoelectric focusing, an electrophoretic method for separating proteins on the basis of their differing isopotential points (see chapter 4).

IgG–Immunoglobulin G, a class of immunoglobulins or antibodies commonly employed in immuno-detection of protein antigens (see chapter 5).

K–One letter abbreviation for the amino acid lysine.

KPi–Potassium phosphate, a buffer component commonly used in preparing protein solutions.

L or Leu–One and three letter abbreviations for the amino acid leucine.

λ–Lambda, symbol for wavelength in optical spectroscopy. Also sometimes used as an abbreviation for microliters (μl).

Lys–Three letter abbreviation for the amino acid lysine.

M or Met–One and three letter abbreviations for the amino acid methionine.

MCE–Mercaptoethanol, a disulfide bond reducing agent.

N–One letter abbreviation for the amino acid asparagine.

N-linked–Referring to glycosylation of a protein by covalent modification of asparagine residues.

nm–Nanometers, a unit of distance equal to 10^{-9} meters used in optical spectroscopy as a unit of wavelength (see chapter 9).

OD–Optical density, another name for absorbance (see chapter 9).

O-linked–Referring to glycosylation of a protein by covalent modification for the hydroxyl groups of serines and/or threonines.

P or Pro–One and three letter abbreviations for the amino acid proline.

PAGE–Polyacrylamide gel electrophoresis (see SDS-PAGE below).

PBS–Phosphate buffered saline, a commonly used buffer for preparing protein solutions.

PEG–Polyethylene glycol.

Phe–Three letter abbreviation for the amino acid phenylalanine.

pI–Isoelectric or isopotential point. The pH at which a molecule has a net charge of zero, and thus does not migrate under the influence of an electric field (see chapter 4).

PITC–Phenylisothiocyanate, a chemical modifying reagent used in amino acid analysis (see chapter 7).

PMSF–Phenylmethylsulphonyl fluoride, a protease inhibitor (see chapter 2).

PTH–Phenylthiohydantoin, chemically modified form of amino acids used for detection in amino acid analysis (see chapter 7).

PTC–Phenylthiocarbamyl, chemically modified form of amino acids used for detection in amino acid analysis (see chapter 7).

Q–The one letter abbreviation for the amino acid glutamine.

R–One letter abbreviation for the amino acid arginine.

R_f–Relative mobility, a measure of the relative migration rate of a molecule used in gel electrophoresis, thin layer chromatography, and size exclusion chromatography (see chapter 4).

S or Ser–One and three letter abbreviations for the amino acid serine.

SDS–Sodium dodecyl sulfate, a detergent used to solubilize proteins and for some electrophoretic methods.

SDS-PAGE–Sodium dodecyl sulfate polyacrylamide gel electrophoresis, a commonly used method for separating proteins on the basis of their differing molecular weights (see chapter 4).

S–S–Disulfide bond.

T or Thr–One and three letter abbreviations for the amino acid threonine.

TBS–Tris buffered saline, a commonly used buffer system for protein solutions.

TCA–Trichloroacetic acid, a reagent used to precipitate proteins from solution (see chapter 2).

TEMED–N,N,N′,N′ tetramethylethylenediamine, a reagent used to prepare polyacrylamide gels (see chapter 4).

TLC–Thin layer chromatography.

T_m–Thermal transition temperature, the temperature at which half of the protein molecules in solution are denatured (see chapter 10).

Trp–Three letter abbreviation for the amino acid tryptophan.

Tyr–Three letter abbreviation for the amino acid tyrosine.

UV–Ultraviolet, pertaining to the portion of the electromagnetic spectrum below ca. 350 nm.

V or Val–One and three letter abbreviations for the amino acid valine.

Vis–Visible, pertaining to the portion of the electromagnetic spectrum between ca. 350 and 900 nm.

W–One letter abbreviation for the amino acid tryptophan.

Y–One letter abbreviation for the amino acid tyrosine.

Appendix II
Suppliers of Reagents and Equipment for Protein Science

Chemicals, Reagents and Proteins

Aldrich Chemical Company, Inc.
940 West Saint Paul Ave.
Milwaukee, WI 53233
(800) 558-9160

Amersham Corporation
2636 South Clearbrook Drive
Arlington Heights, IL 60005
(800) 323-9750

Bio-Rad Laboratories
1414 Harbour Way S.
Richmond, CA 94804
(800) 426-6723

Boehringer Mannheim Corporation
Biochemical Products
9115 Hague Road, P.O. Box 50414
Indianapolis, IN 46250-0414
(800) 262-1640

Calbiochem
P.O. Box 12087
San Diego, CA 92112
(800) 854-9256

Eastman Kodak Co.
343 State St.
Building 701
Rochester, NY 14650
(800) 225-5352

Pharmacia LKB Biotechnology AB
800 Centennial Ave.
Piscataway, NJ 08854
(800) 526-3618

Pierce Chemical Co.
P.O. Box 117
Rockford, IL 61105
(800) 874-3723

Sigma Chemical Co.
P.O. Box 14508
St. Louis, MO 63178
(800) 325-3010

United States Biochemical Corp.
P.O. Box 22400
Cleveland, Ohio 44122
(800) 321-9322

Equipment (General)

Amicon
24 Cherry Hill Drive
Danvers, MA 01923
(800) 343-1397

Beckman Instruments, Inc.
P.O. Box 3100
Fullerton, CA 92634-3100
(800) 742-2345

Bio-Rad Laboratories
1414 Harbour Way S.
Richmond, CA 94804
(800) 426-6723

B. Braun Instruments
824 12th Street
Bethlehem, PA 18018
(800) 258-9000

Hamilton Instruments
P.O. Box 100030
Reno, NV 89520
(702) 786-7077

Hoefer Scientific Instruments
P.O. Box 77387
654 Minnesota St.
San Francisco, CA 94107
(800) 227-4750

Millipore Corp.
80 Ashby Road
Bedford, MA 01730
(800) 225-1380

Novex
4202 Sorrento Valley Blvd.
San Diego, CA 92121
(800) 456-6839

Pharmacia LKB Biotechnology AB
800 Centennial Ave.
Piscataway, NJ 08854
(800) 526-3618

Schleicher & Schuell, Inc.
10 Optical Ave.
Keene, NH 03431
(800) 245-4024

Spectrum Medical Industries, Inc.
1100 Rankin Road
Houston, Texas 77073-4716
(800) 634-3300

Equipment (Spectroscopic)

Beckman Instruments, Inc.
P.O. Box 3100
Fullerton, CA 92634-3100
(800) 742-2345

Hitachi Instruments, Inc.
Chicago Technology Park
Suite 315
2201 West Campbell Park Drive
Chicago, IL 60612
(800) 548-9001

Jasco, Inc.
8649 Commerce Drive
Easton, Maryland 21601
(800) 333-5272

Olis
130 Conway Drive Suites A & B
Bogart, Georgia 30622-1724
(800) 852-3504

Photon Technology International, Inc.
1 Deerpark Drive, Suite F
South Brunswick, NJ 08852
(908) 329-0910

Spex Industries, Inc.
3880 Park Ave.
Edison, NJ 08820
(800) 522-7739

Index

MIX
Papier aus verantwortungsvollen Quellen
Paper from responsible sources
FSC® C105338

FSC
www.fsc.org

If you have any concerns about our products,
you can contact us on
ProductSafety@springernature.com

In case Publisher is established outside the EU,
the EU authorized representative is:
Springer Nature Customer Service Center GmbH
Europaplatz 3, 69115 Heidelberg, Germany

Printed by Libri Plureos GmbH
in Hamburg, Germany